PHYSICS FORMULAS AND TABLES OF CONSTANTS

Formulas and Physical Constants for Students and Teachers

Kingsley Augustine

Table of Contents

PART 1: FORMULAS

Fundamental quantities and units

Quantity	Unit	Abbreviation
Length	Meter	m
Time	Seconds	S
Mass	Kilogram	Kg
Electric Current	Ampere	A
Thermodynamic temperature	Kelvin	K
Amount of substance	Mole	Mol
Luminous intensity	Candela	Cd

Friction

Frictional force is given by:

$F = \mu R$,

where μ is a constant called the coefficient of friction. The normal reaction, R, is given by:

$R = mg$,

where m is the mass of the body, and g is the acceleration due to gravity.

Friction on inclined planes

When a body of mass, m, moves down an inclined plane, with an angle of inclination, θ, the force with which it moves downward is given by:

$F = mg\sin\theta$ (This is when the plane is smooth and there is no friction)

g is acceleration due to gravity.

When there is friction on the plane (rough plane), the frictional force holding the body on the plane is also given by:

$$F = mg\sin\theta$$

Also, $F = uR$ (force of friction)

where $F = mg\sin\theta$, $R = mg\cos\theta$, and μ = coefficient of friction.

$$\mu = \frac{mg\sin\theta}{mg\cos\theta}$$

$$\mu = \tan\theta$$

Note that $mg\sin\theta$ is the force parallel to the plane, while $mg\cos\theta$ is the force perpendicular to the plane.

Viscosity
Terminal velocity
When a ball falls through a viscous liquid the equation of motion is given by:

$$mg - V - U = ma$$

where V = viscous force and U = upthrust
At terminal velocity:

$$V = mg - U$$

If at terminal velocity, the vertical distance moved by a falling ball at time, t, is determined, then the terminal velocity, v_{tm}, of the ball can be calculated as follows:

$$v_{tm} = \frac{L}{t}$$

where L is the distance moved by the ball.

Density

$$\text{Density} = \frac{\text{mass}}{\text{volume}}$$

Or, $\quad p = \dfrac{m}{V}$

Relative density

$$\text{Relative density (R.D)} = \frac{\text{Density of substance}}{\text{Density of water}}$$

If we assume constant volume, then:

$$\text{Relative density (R.D)} = \frac{\text{mass of substance}}{\text{mass of an equal volume of water}}$$

Or (in terms of weight),

$$\text{Relative density (R.D)} = \frac{\text{weight of substance}}{\text{weight of an equal volume of water}}$$

I f a solid weighs w_1 in air, w_2 when immersed in water and w_3 when immersed in a liquid of unknown density, then the following formulas apply:

$$\text{Relative density of solid (R.D)}_{solid} = \frac{w_1}{w_1 - w_2}$$

Which means, $\quad \dfrac{\text{weight of solid in air}}{\text{Loss of weight in water}}$

$$\text{Relative density of liquid (R.D)}_{liquid} = \frac{w_1 - w_3}{w_1 - w_2}$$

which means, $\dfrac{\text{Loss of weight in liquid}}{\text{Loss of weight in water}}$

Pressure

$$\text{Pressure} = \frac{\text{Force}}{\text{Area}}$$

Or, $P = \dfrac{F}{A}$

Or, $P = \dfrac{W}{A}$

where, F = force, A = area, and W = weight of the object

Pressure in liquids

The force which act normally to this surface of a liquid at rest is called fluid pressure. The formula for pressure in fluid is given by:

$$P = h\rho g,$$

where P = pressure, h = height or depth in the liquid, ρ = density of the liquid, and g = acceleration due to gravity.

However, at a particular point in a liquid in an open container, the total pressure acting is given by:

$$P = H + h\rho g,$$

where H is the atmospheric pressure.

Work, Energy and Power

Work

Work done = Force x distance moved

Or, $W = F \times d$

When a non-parallel force, F, which makes an angle θ to the horizontal acts on a body and moves the body through a distance, d, then the work done is given by:

$$W = F\cos\theta \times d$$

Mechanical energy

1. Potential energy
$$P.E = mgh$$
Where P.E is the potential energy, m is the mass of the body in kg, g is the acceleration due to gravity, h is the vertical height of the body above the ground in m.

2. Kinetic energy, K.E

$$K.E = \tfrac{1}{2}mv^2$$

where m is mass in kg, and v is the velocity of the body in m/s.

Conservation law of mechanical energy
$$P.E + K.E = Constant$$

i.e. P.E + K.E at any point = P.E + K.E at another point.

Power

$$Power = \frac{Work\ or\ energy\ used}{Time\ taken}$$

i.e. $Power = \dfrac{W}{t}$

Also, Power = f x v

1h.p = 746W = 0.746kW
h.p = horse power

Current Electricity

Electric Current

If a quantity of charge, Q columns flows for t seconds, the current, I, is given by:

$$I = \frac{Q}{t}$$

Ohm's Law

$$V = IR$$

where I is the current and R is a constant called resistance.

Electromotive force (E.M.F)

$$E = I(R + r)$$

Where E = emf, I is the current in the circuit, R is the effective/combined external resistance, and r is the effective/combined internal resistance.

Or, $E = V + Ir$

Lost volt = Ir

Resistivity

$$\rho = \frac{RA}{l}$$

where ρ is the resistivity, A is the cross-sectional area and l = length of the wire/conductor.

Conductivity

The reciprocal of resistivity gives the conductivity, G, of a wire/conductor. Conductivity is given by:

$$G = \frac{1}{\rho},$$

Or, $G = \dfrac{l}{RA}$

Resistors in Series

When n resistors of resistance R_1, R_2, R_3, ..., R_n are connected in series, their effective or total or combined resistance is given by:

$$R = R_1 + R_2 + R_3 + ... + R_n$$

Resistors in Parallel

When n resistors of resistance R_1, R_2, R_3, ..., R_n are connected in parallel, their effective or total or combined resistance is given by:

$$\frac{1}{R} = \frac{1}{R_1} + \frac{1}{R_2} + \frac{1}{R_3} + ... + \frac{1}{R_n}$$

The inverse of $\frac{1}{R}$, gives the effective resistance R.

Cells in Series

When n cells each of e.m.f E_1, E_2, E_3, ..., En are connected in series, their effective or total or combined emf is given by:

$$E = E_1 + E_2 + E_3 .... + E_n$$

The effective internal resistance of the cells is given by:

$$r = r_1 + r_2 + r_3 + ... + r_n$$

Cells in Parallel

When n cells each of e.m.f E_1, E_2, E_3, ..., E_n are connected in parallel, their effective or total or combined emf is given by:

$$E = E_1 \qquad \text{(any one of the emfs of the cells)}$$

The effective internal resistance of the cells is given by:

$$\frac{1}{r} = \frac{1}{r_1} + \frac{1}{r_2} + \frac{1}{r_3} + ... + \frac{1}{r_n}$$

Therefore, r = the reciprocal of $\frac{1}{r}$

Also, r can be given by: $r = \frac{r_1}{n}$, where n is the total number of cells.

Wheatstone Bridge

$$\frac{R_1}{R_2} = \frac{R_3}{R_4}$$

Meter Bridge

$$\frac{R_1}{R_2} = \frac{l_1}{l_2}$$

Where l_1 is the balance point from the left hand side of the meter bridge. Since the wire in the meter bridge is 1 meter (100cm) long, then, $l_2 = 100 - l_1$

Comparing e.m.f of cells by using potentiometer

The expression used for this comparison is given by:

$$\frac{E_1}{E_2} = \frac{l_1}{l_2}$$

The emf to be determined is related to the voltage across the potentiometer wire as follows:

$$\frac{V}{E} = \frac{l}{100}$$

where E is the emf to be determined, V is the voltage across the potentiometer wire, and l is the balance length. 100 represent the length of the potentiometer wire which is usually 100cm.

Conversion of a galvanometer to an ammeter

To convert a galvanometer to an ammeter a low resistance shunt, R_S, is connected in parallel with a suitable galvanometer. If the resistance of the galvanometer is R_G, and the current through it is I_G, then the equation below is valid:

$$IR_S = I_G(R_S + R_G),$$

where I is the current that the galvanometer is to be converted to read. From the equation above, R_S is given by:

$$R_S = \frac{I_G R_G}{I - I_G}$$

Conversion of a galvanometer to a voltmeter

To convert a galvanometer to voltmeter, a high resistance multiplier, R_m is joined in series with a suitable galvanometer. If the resistance of the galvanometer is R_G, and the current through it is I_G, then the p.d or voltage, V, cross the combination is given by:

$$V = I_G(R_m + R_G)$$

where V is the voltage that the galvanometer is to be converted to read.

From the equation above, R_m, is given by:

$$R_m = \frac{V}{I_G} - R_G$$

Electrical Energy

If the pd applied is V volts and the quantity of electricity is Q coulombs, the work done is given by:

$$W = VQ$$

Or, $W = VIt$ (Since $Q = It$)

Hence, electrical energy is given by:

$$E = IVt$$ (The time should be in seconds)

Or, $E = I^2Rt$ (Since $V = IR$)

Or, $E = \dfrac{V^2t}{R}$ (Since $I = \dfrac{V}{R}$)

Electrical Power

$$Power = \frac{Electrical \ \ energy}{time}$$

$$P = \frac{E}{t}$$

14

$$P = IV$$

Or, $$P = I^2R$$

Or, $$P = \frac{V^2}{R}$$

Buying of Electrical Energy

1KWh = 1000W for 1 hour

Elasticity of material

Hooke's Law

$$F = Ke$$

where F = applied force, e = extension produced, and K = elastic constant.

Young's modulus

Suppose a force, F (in N), is applied to a wire of cross sectional area, A (in m²), and original length, L (in m), and this results in an extension, e (in m), then we can define the parameters of stress and strain on the the wire as follows:

$$\text{Stress} = \frac{\text{Force}}{\text{Area}}$$

$$\text{Stress} = \frac{F}{A}$$

$$\text{Strain} = \frac{\text{Extension}}{\text{Original length}}$$

$$\text{Strain} = \frac{e}{L}$$

$$\text{Young's modulus, } Y = \frac{\text{Stress}}{\text{Strain}}$$

Energy Stored in an Elastic material

Workdone = Average force x extension

$W = \frac{1}{2}Fe$

Energy stored, $E = \frac{1}{2}Fe$

Or, $E = \frac{1}{2}Fe$

Or, $E = \frac{1}{2}Ke^2$

Therefore, $E = W = \frac{1}{2}Fe$

Or, $E = \frac{1}{2}Ke^2$

Elastic Potential Energy

$E = \frac{1}{2}Fe$

Or, $E = \frac{1}{2}Ke^2$

Resolution of vectors

Let us use a vector such as force, F, to state our formulas as given below.

$F_x = F\cos\theta$

where F_x is the horizontal component of F, and θ is the angle that F makes with the positive x-axis.

For a vector of magnitude, F, the vertical component of the vector is given by:

$Fy = F\sin\theta$

where Fy is the vertical component of F, and θ is the angle that F makes with the positive x-axis.

Motion

The force on a body in circular motion

For a body performing circular motion, the linear velocity of the body is given by:

$V = \omega r,$

where ω is the angular velocity in rad/sec, while r is the radius of the circular path. Its acceleration is given by:

$$a = \omega^2 r$$

Or $a = \dfrac{v^2}{r}$

Since, F = ma, therefore the force acting on the body is given by:

$$F = \dfrac{mv^2}{r}$$

Speed of a body in linear motion

$$\text{Speed} = \dfrac{\text{Distance}}{\text{time}}$$

Average speed

$$\text{Average speed} = \dfrac{\text{Total Distance covered}}{\text{Total time taken}}$$

Velocity

$$\text{Velocity} = \dfrac{\text{Distance moved in a stated direction}}{\text{time taken}}$$

Or,

$$\text{Velocity} = \dfrac{\text{Displacement}}{\text{time}}$$

Acceleration

$$\text{Acceleration} = \dfrac{\text{Change in velocity}}{\text{time taken for the change}}$$

Uniform or constant acceleration

Acceleration, a, is given by:

$$a = \dfrac{v - u}{t}$$

where v is the final velocity, u is the initial velocity and t is the time.

When the rate of change of velocity of a body is not always constant, the body is said to have non-uniform acceleration.

Position of a body in the x, y plane

The distance between two points $A(x_1, y_1)$ and $B(x_2, y_2)$ is given by:

$$AB = \sqrt{(x_2 - x_1)^2 + (y_2 - y_1)^2}$$

Equations of motion

Recall that if a body is moving at a constant velocity, v, the distance, s, covered in time, t, is given by:

$$s = vt \quad \left(\text{since } v = \frac{s}{t}\right)$$

If the body was already having an initial velocity, u, it means that, v, becomes the final velocity after it has accelerated at a constant value, a, in time, t, and covered a distance, s. Then the four equations of motion are given by:

1. $s = \dfrac{(u + v)t}{2}$

2. $v = u + at$

3. $s = ut + \frac{1}{2}at^2$

4. $v^2 = u^2 + 2as$

Note that when a body starts from rest, its initial velocity, u = 0, and when a body comes to rest, its final velocity, v = 0. Note also that retardation is negative acceleration, that is, -a.

Motion under gravity

The equations of motion for a body falling under gravity is given by replacing a = g in the equation of linear motion. This gives the equations below:

1. $v = u + gt$

2. $s = ut + \frac{1}{2}gt^2$

3. $v^2 = u^2 + 2gs$

Similarly, when a body is projected vertically upwards, its motion is obtained by replacing g = −g. This gives the equations below:

18

1. $v = u - gt$
2. $s = ut - \frac{1}{2}gt^2$
3. $v^2 = u^2 - 2gs$

When a body is released from a height above the ground, such a body falls freely, with u = 0. When a body is thrown upwards, its final velocity, v, at the highest point is, v = 0

However, when a body is thrown horizontally with an initial velocity, u, from a height, h, above the ground, its vertical motion is assumed to be a free fall. Hence, its equation of motion is given by:

$h = \frac{1}{2}gt^2$

For such a body, its horizontal distance covered is given by:

$s = ut$

Projectiles
When a body is projected upwards with an initial velocity, u, at an angle, θ, its vertical velocity at any point in time, t, is given by:

$v = u\sin\theta - gt$

It horizontal velocity, v_x is given by:

$v_x = u\cos\theta$ (This value is constant throughout the motion)

Time of flight

$$t = \frac{u\sin\theta}{g}$$

where t = the time taken to reach the maximum height.
The total time of flight, T, is twice this time taken to reach the maximum height. It is given by:

$$T = \frac{2u\sin\theta}{g}$$

Maximum height

The maximum height attained by a projectile is given by:

$$H = \frac{u^2 \sin^2 \theta}{2g}$$

Or, $\quad H = \frac{(u\sin\theta)^2}{2g}$

The vertical height attained by a projectile is given by:

$$H = ut\sin\theta - \tfrac{1}{2}gt^2$$

Range

The horizontal distance covered by a projectile is given by:

$$s = ut\cos\theta$$

The range (total horizontal distance) of a projectile is given by:

$$R = \frac{u^2 \sin 2\theta}{g}$$

The maximum range is obtained when $\sin 2\theta = 1$, if the body is projected at $\theta = 45$.

Hence maximum range is given by:

$$R_{max} = \frac{u^2}{g}$$

Momentum

Momentum = mass x velocity

$$= mv$$

The force acting on a moving body of mass, m, and acceleration, a, is given by:

$$F = ma$$

Or, $\quad F = \dfrac{m(v - u)}{t} \quad$ (since $a = \dfrac{v - u}{t}$)

Therefore, $\quad Ft = m(v - u)$

Or, $\quad Ft = mv \quad$ (when the initial velocity, $u = 0$)

The quantity, Ft, is known as the impulse of the body. Hence, impulse is equal to momentum and are related by:

Recoil velocity

When a bullet is fired out of a gun, the total momentum of the gun and bullet are related by:

$$m_b \times v_b = m_g \times v_g$$

where b represents the bullet and g represents the gun.

Principle of conservation of linear momentum

$$m_1 u_1 + m_2 u_2 = m_1 v_1 + m_2 v_2$$

where m_1 and m_2 are the masses of the two bodies. u_1 and u_2 are their initial velocities, while v_1 and v_2 are their final velocities.

Types of collision

1. **Elastic collision:**

Kinetic energy before collision = kinetic energy after collision

Thus, $\quad \frac{1}{2}(m_1 u_1{}^2 + m_2 u_2{}^2) = \frac{1}{2}(m_1 v_1{}^2 + m_2 v_2{}^2)$

Or, $\quad (m_1 u_1{}^2 + m_2 u_2{}^2) = (m_1 v_1{}^2 + m_2 v_2{}^2)$

2. **Inelastic collision:**

Loss of energy $= \frac{1}{2}(m_1 u_1{}^2 + m_2 u_2{}^2) - \frac{1}{2}(m_1 v_1{}^2 + m_2 v_2{}^2)$

Weight of a body in a lift

1. When a body of mass, m, is in a lift, and the lift is accelerating upwards, the equation of motion for such a situation is:

$$R = m(g + a)$$

where, R is the reaction exacted by the floor of the lift, a, is the acceleration of the lift, while g is the acceleration due to gravity.

2. When the lift is accelerating downwards the equation for this case is:

$$R = m(g - a)$$

3. When the lift is moving at a constant speed, then, $R = mg$

Simple Harmonic Notion

Velocity of simple harmonic motion

$$v = \omega A$$

where, ω = angular velocity in rad/sec, and A = maximum displacement of the body from its fixed point (centre of motion)

However, at a point of distance, x, from its fixed points, where x is less than A, the velocity is given by:

$$v = \omega\sqrt{A^2 - x^2}$$

Acceleration of simple harmonic motion

The acceleration, a, of a body in simple harmonic motion is given by:

$$a = \omega^2 x$$

It should be noted that this acceleration is always negative.

Period, frequency and amplitude of simple harmonic motion

The period of a body performing simple harmonic motion is generally expressed as follows:

$$T = \frac{2\pi}{\omega} \quad (\omega = \text{angular velocity})$$

Or, $T = \dfrac{1}{f}$

where, f, is the frequency of oscillation.

The angular velocity is given by:

$\omega = 2\pi f$,

For a simple pendulum, the period Is given by:

$$T = 2\pi \sqrt{\dfrac{l}{g}}$$

where, l is the length of the pendulum, and g is acceleration due to gravity.

For a particle of mass, m, suspended by a spiral spring, the period is given by:

$$T = 2\pi \sqrt{\dfrac{m}{k}}$$

where k = elastic constant of the spring.

Or, $T = 2\pi \sqrt{\dfrac{e}{g}}$

where, e = extension of the spring.

For a loaded test tube oscillating in a liquid, if its mass is m, cross sectional area is A, and density of the liquid is p, then its period is given by:

$$T = 2\pi \sqrt{\dfrac{m}{Apg}}$$

Energy of simple harmonic motion

Potential energy, $P.E_{max} = \frac{1}{2}kx^2$

where $k = \dfrac{F}{e}$ (from Hooke's law), and it is the force constant of the spring.

x is the distance through which the string was stretched before it was displaced.

The kinetic energy of the motion is given by:

$$K.E_{max} = \frac{1}{2}mv^2 \qquad \text{(where } v = wx)$$

Machines

Mechanical advantage (M.A)

$$M.A = \frac{\text{Load}}{\text{Effort}}$$

Velocity ratio (V.R)

$$V.R = \frac{\text{Effort distance}}{\text{Load distance}}$$

Efficiency (E)

$$E = \frac{\text{work output}}{\text{work input}} \times 100$$

where, work output = load x load distance, and work input = effort x effort distance

Or, $\quad E = \dfrac{M.A}{V.R} \times 100$

Pulleys

V.R = Total number of pulleys

Note that, one fixed pulley system has a velocity ratio, V.R = 1, while one movable pulley system, has velocity ratio, V.R = 2.

Inclined plane

The velocity ratio of an inclined plane is given by:

$$V.R = \frac{l}{h}$$

where l = the length of the hypotenuse side of the right angled triangle formed by the plane, while h is the opposite side, which is the vertical height of the plane above the ground.

Or, $V.R = \dfrac{1}{\sin \theta}$

where is the angle that the plane makes with the horizontal

Screw jack
The velocity ratio of a screw jack is given by:

$$V.R = \dfrac{2\pi a}{\text{pitch}}$$

where a = radius of the arm of rotation of the handle, and pitch = distance between two consecutive threads of the screw jack.

Wheel and axle
The velocity ratio of a wheel and axle is given by:

$$V.R = \dfrac{R}{r}$$

where, R = radius of the wheel, while r = radius of the axle

Gear wheels
The velocity ratio of a gear system is given by:

$$V.R = \dfrac{\text{number of teeth of the driven (larger) wheel}}{\text{number of teeth of the driver (smaller) wheel}}$$

Hydraulic press
The velocity ratio of a hydraulic press is given by:

$$V.R = \dfrac{F_2}{F_1}$$

Or, $V.R = \dfrac{A_2}{A_1}$

where F_2 = force on load piston, F_1 = force on efforts piston, A_2 = area of load piston, A_1 = area of efforts piston.

Heat

Temperature scale

In order to convert from Fahrenheit to Celsius, we use the formula:

$$(F - 32) \times \frac{5}{9}$$

where F is the temperature in Fahrenheit.

In order to convert from Celsius to Fahrenheit, we use the formula:

$$\left(C \times \frac{9}{5}\right) + 32$$

where C is the temperature in Celsius

Linear Expansivity

$$\alpha = \frac{L_2 - L_1}{L_1(\theta_2 - \theta_1)}$$

where α is the linear expansivity, L_2 = final length of the body, L_1 = initial length of the body, θ_2 = final temperature of the body, θ_1 = initial temperature of the body.

A condensed formula for linear expansivity is:

$$\alpha = \frac{L}{L_1\theta}$$

where L = change in length of the body, L_1 = original length of the body, θ = change in temperature.

Area expansivity

Area expansivity, $\beta = \frac{A_2 - A_1}{A_1(\theta_2 - \theta_1)}$

Or, $\beta = \frac{A}{A_1\theta}$

where A = change in area, and 1 and 2 means initial and final values

Volume expansivity

Volume or cubic expansivity is given by:

$$\gamma = \frac{V_2 - V_1}{V_1(\theta_2 - \theta_1)}$$

Or, $\quad \gamma = \dfrac{V}{V_1\theta}$

where V = change in volume, V_1 = original volume, $\theta = \theta_2 - \theta_1$.

Area and volume expansivity are related to linear expansivity by the following expressions:

Area expansivity, $\beta = 2\alpha$,

Volume expansivity, $\gamma = 3\alpha$,

where α = linear expansivity

Real and apparent expansivity

The real (absolute) cubic expansivity (γ_r) of a liquid and the apparent cubic expansivity (γ_a) of a liquid are related by:

$$\gamma_r = \gamma_a + \gamma$$

where γ is the cubic or volume expansivity of the vessel.
The apparent cubic expansivity, γ_a, is given by:

$$\gamma_a = \frac{\text{Volume or mass of liquid expelled}}{\text{Volume or mass of liquid remaining} \times \text{temperature rise}}$$

Heat Energy

The heat energy received or given out when there is a change in the temperature of a body is given by:

$$Q = mc\Delta\theta$$

Or, $\quad Q = mc(\theta_2 - \theta_1)$

where Q = quantity of heat, m = mass, c = specific heat capacity, and $\Delta\theta$ = $(\theta_2 - \theta_1)$ = change in temperature, θ_2 = higher temperature, while θ_1 = the lower temperature.

Heat capacity

Heat capacity is given by the expression:

$$H = mc$$

where H = heat capacity, m = mass in kg and c = specific heat capacity.

Electrical Heat Energy

When an electrical energy is supplied to a body, the following equations can be applied:

$$IVt = mc\Delta\theta$$

Or, $\quad I^2Rt = mc\Delta\theta$

Or, $\quad \dfrac{V^2t}{R} = mc\Delta\theta$

Change of state and Latent heat

The heat energy needed to change the state of a body is given by:

$$Q = mL$$

where Q = quantity of heat, m = mass, and L = the specific latent heat.

Humidity

Relative humidity, R.H, is defined as a ratio as follows:

$$R.H = \frac{\text{mass of water vapour present in a certain amount of air}}{\substack{\text{mass of water vapour required to saturate} \\ \text{the same volume of air at the same temperature}}} \times 100$$

Or,

$$R.H = \frac{\text{saturated vapour pressure of water at the dew point}}{\substack{\text{saturated vapour pressure of water} \\ \text{at the air temperature}}} \times 100$$

Gas Laws

Boyle's law

$$V = \frac{k}{P} \quad \text{(constant temperature)}$$

where V = volume, P = pressure, and k = constant.

Therefore, $PV = k$

If the volume or pressure of a gas changes from an initial value to a final value, then:

$$P_1V_1 = P_2V_2$$

Practical application of Boyle's law

When a capillary tube is vertical, with the open end facing upwards, the pressure and volume of the air in the tube can be determined as follows:

$P = H + h$, and $V = l$

where P = pressure of air in the tube, H = atmospheric pressure, h = length of mercury in the tube, V = volume of trapped air in the tube, and l = length of trapped air in the tube.

When the tube is vertical with the open end facing downwards, the pressure and volume of the air in the tube can be determined as follows:

$P = H - h$, and $V = l$

When the tube is horizontal, the following applies:

$P = H$, and $V = l$

Charles law

$$V = kT$$

where, V = volume, T = temperature in kelvin, and k = constant.

Therefore:

$$\frac{V}{T} = k$$

If the volume or temperature of a gas changes from an initial value to a final value, then:

$$\frac{V_1}{T_1} = \frac{V_2}{T_2}$$

Pressure law

$$P = KT,$$

Or $\quad \dfrac{P_1}{T_1} = \dfrac{P_2}{T_2}$

The general gas law

$$\frac{PV}{T} = k$$

Or, $\quad \dfrac{P_1 V_1}{T_1} = \dfrac{P_2 V_2}{T_2}$

Wave Motion

Period, T, is expressed as:

$$T = \frac{1}{f}$$

Or, $\quad T = \dfrac{2\pi}{\omega}$

Or, $\quad T = \dfrac{\text{time taken for oscillation}}{\text{number of oscillation}}$

where ω = the angular velocity and f = frequency.

Wave velocity is given by:

$$v = \frac{\lambda}{T}.$$

Or, $\quad v = f\lambda$

where T = period and λ = wavelength

Mathematical representation of wave motion

The displacement of a wave at any point can be expressed as:

$$y = A \sin\theta$$

where θ = phase angle, A = amplitude and y is the vertical distance.

Also,

$$y = A \sin\left(\omega t - \frac{\omega x}{v}\right)$$

where ω = angular velocity and x = horizontal distance.

The displacement y, of a wave can also be represented by:

$$y = A \sin\left(2\pi ft - \frac{2\pi fx}{v}\right)$$

Or, $$y = A \sin\left(2\pi ft - \frac{2\pi x}{\lambda}\right)$$

Beats

Beat frequency is given by:

$$f = f_1 - f_2$$

Sound

Applications of echoes

$$v = \frac{2x}{t}$$

where, x is the distance in metres between the source of sound and the reflecting surface, while t, is the time in seconds taken to hear the echo.

Vibrations in a closed pipe

At first resonance, the length of the air column is given by:

31

$$l = \frac{\lambda}{4} \quad \text{(where } \lambda = \text{wavelength)}$$

The associated frequency of vibration at first resonance is given by:

$$f_o = \frac{v}{\lambda} = \frac{v}{4l}$$

where v, is the speed of sound in air, f_o is the fundamental frequency of air in the closed pipe.

The second resonance is observed at:

$$l = \frac{3\lambda}{4}$$

In this case the frequency of vibration is given by:

$$f_1 = \frac{3v}{4l} = 3f_o$$

This frequency, $f_1 = 3f_o$, is called the third harmonic or first overtone of a closed pipe (f_o is the first harmonic)

Vibration in an open pipe

The length of air column at first resonance is given by:

$$l = \frac{\lambda}{2}$$

The fundamental frequency in this case is therefore, $f_o = \frac{v}{2l}$

The next harmonic occurs at, $l = \lambda$.

The frequency of vibration at this point is given by:

$$f_1 = \frac{v}{l}$$

$$f_1 = 2f_o$$

Modes of vibration of a stretched string

$$l = \frac{\lambda}{2}$$

$$f_0 = \frac{v}{2L}$$

The velocity of a wave propagated along a fixed wire or string is given by:

$$v = \sqrt{\frac{T}{m}}$$

The fundamental frequency can therefore be given by:

$$f_0 = \frac{v}{2l}$$

$$= \frac{1}{2l}\sqrt{\frac{T}{m}}$$

where T is the tension in the string, and m is the mass per unit length of the sting.

Note that f is:

directly proportional to $\sqrt{T}$ $(f\,\alpha\,\sqrt{T})$

inversely proportional to l $(f\,\alpha\,\frac{1}{l})$

inversely proportional to $\sqrt{m}$ $(f\,\alpha\,\frac{1}{\sqrt{m}})$

Light

Pin Hole Camera
Magnification produced by the pinhole camera

Magnification, $M = \dfrac{\text{size of image}}{\text{size of object}}$

Or, $M = \dfrac{\text{height of image}}{\text{height of object}}$

Or, $M = \dfrac{\text{Length of camera}}{\text{distance of object from pin hole}}$

Images formed by inclined mirrors

When two plane mirrors are placed at an angle to each other, the number of images formed by the plane mirrors is given by:

$$n = \left(\frac{360}{\theta}\right) - 1$$

where n is the number of images, θ is the angle between the two mirrors.

Rotation of plane mirrors

If the direction of an incident ray on a mirror is kept constant and the mirror is rotated through an angle, it will be found that the reflected ray will be rotated through twice that angle.

This means that the angle of rotation of the reflected ray when a mirror is rotated through an angle θ, is given by, 2θ.

Curved mirrors

The mirror formula is given by:

$$\frac{1}{f} = \frac{1}{u} + \frac{1}{v}$$

where u = object distance, v = image distance and f = focal length of the mirror.

Refractive index

When light travel from a first medium to a second medium, the refractive index of the material of the second medium is given by:

$$n = \frac{\sin i}{\sin r}$$

where n = refractive index of the second medium with respect to the first medium, i = angle of incidence in the first medium, and r = angle of refraction in the second medium. If light is travelling from air to glass, the refractive index of glass is given by:

$$_a n_g = \frac{\sin i \ (\text{in air})}{\sin r \ (\text{in glass})}$$

Similarly, if light travels from glass to air, then the refractive index of air is given by:

$$_g n_a = \frac{1}{_a n_g}$$

Since refraction depends on the change in speed of light from one medium to another, refractive index is also given by:

$$_a n_g = \frac{\text{speed of light in air (vacuum)}}{\text{speed of light in glass}}$$

In terms of wavelength refractive index can also be given by:

$$_a n_g = \frac{\text{wavelength of light in air}}{\text{wavelength of light in glass}}$$

Refraction of light through triangular prisms

At minimum deviation, the refractive index, n, of a glass, the refracting

35

angle, A, of the prism and the angle of minimum deviation, D_m, are related by the equation:

$$n = \frac{\sin\left(\frac{A + D_m}{2}\right)}{\sin\left(\frac{A}{2}\right)}$$

Note that the refracting angle (A), of a triangular glass prism is the vertical angle of the triangle.

Critical angle and refractive index

The critical angle of a medium and its refractive index are related by the expression:

$$n = \frac{1}{\sin c}$$

where, c, is the critical angle of the denser medium when light travels from a denser medium to a less dense medium.

Real and apparent depth

Real depth, apparent depth and refractive index are related by the formula:

$$n = \frac{\text{real depth}}{\text{apparent depth}}$$

Use of the lens formula

The lens formula is given by:

$$\frac{1}{f} = \frac{1}{u} + \frac{1}{v}$$

The magnification is given by:

$$M = \frac{\text{Image height}}{\text{object height}}$$

Or, $$M = \frac{\text{Image distance}}{\text{Object distance}}$$

Or, $M = \dfrac{v}{u}$

Optical Instruments

The compound microscope
The magnifying power (magnification) of a compound microscope is given by:

$M = m_1 \times m_2$

where, m_1 = the magnification produced by the objective lens, and m_2 = the magnification produced by the eye-piece. The magnification can also be given by:

$$M = \frac{\text{Height of final image}}{\text{Height of the object}}$$

The astronomical telescope
The magnification of the astronomical telescope is given by:

$$M = \frac{\text{Focal length of the objective lens}}{\text{Focal length of the eyepiece}}$$

The Galilean telescope
The magnification of the Galilean telescope is given by:

$$M = \frac{\text{Focal length of the objective lens}}{\text{Focal length of the eyepiece}}$$

Gravitational Field
Newton's law of universal gravitation
The force of attraction between any two objects in the universe is given by:

$$F = \frac{G m_1 m_2}{r^2}$$

where m_1 and m_2 are the masses of the objects, r is their distance apart, and G is the gravitational constant.

Acceleration due to gravity

The acceleration due to gravity, g, can be calculated as follows:

$$g = \frac{GM}{R^2}$$

where M = mass of the earth, R = radius of the earth, and G = gravitational constant

Gravitational potential energy

The gravitational potential energy at a point in a gravitational field is given by:

$$G_P = -\frac{GMm}{r}$$

Gravitational potential

The gravitational potential at a point in a gravitational field is given by:

$$V = -\frac{GM}{r}$$

Escape velocity

Escape velocity is represented by:

$$v_E = \sqrt{\frac{2GM}{R}}$$

Or, $\quad v_E = \sqrt{2gR}$

where, G = gravitational constant, M = mass of the earth, R = radius of the earth, and g = acceleration due to gravity

Electric field

The force of attraction between two charges in an electric field is given by:

$$F = \frac{q_1 q_2}{4\pi\varepsilon_0 r^2}$$

where q_1 and q_2 are the charges, r is their distance apart, and $\frac{1}{4\pi\varepsilon_0}$ is a constant. ε_0 is a constant

Electric field intensity

Electric field intensity is given by:

$$E = \frac{F}{q}$$

Or, $\quad E = \dfrac{Q}{4\pi E_0 r^2}$

Electric potential

The electric potential, V, at any point in an electric field is given by:

$$V = \frac{Q}{4\pi E_0 r}$$

The work done in an electric field is given by:

$$W = qV$$

where q is the charge, and V is the potential difference (voltage) applied.

The electric field intensity, E, and electric potential difference, V, are related by:

$$E = \frac{V}{d}$$

where d if the distance between the charged plates.

Capacitance of a capacitor

The capacitance, C, of a capacitor is given by:

$$C = \frac{q}{V}$$

where q = charge and v = voltage

For a parallel plate capacitor, the capacitance is given by:

$$C = \frac{\varepsilon A}{d}$$

where A = area of the plates, d = distance between the plates, and ε = a constant call the permittivity

Arrangement of capacitors

1. When capacitors are arranged in series, the effective capacitance is given by:

$$\frac{1}{C} = \frac{1}{C_1} + \frac{1}{C_2} + \frac{1}{C_3} + ...$$

The reciprocal of $\frac{1}{C}$ = C.

2. For a parallel arrangement, the equivalent capacitance is given by:

$$C = C_1 + C_2 + C_3 + C_4 + ...$$

Energy stored in a capacitor

The energy stored in a capacitor is given by:

$$E = \tfrac{1}{2}qV.$$

Or, $E = \tfrac{1}{2}CV^2$

Or, $E = \dfrac{q^2}{2C}$

Electrolysis

Laws of Electrolysis

(1) Faraday's First Law of Electrolysis is expressed mathematically as follows:

$$m = Zit$$

where m = mass of element deposited, I = current, t = time in seconds, and Z is a constant called the electrochemical equivalent of the metal deposited.

(2) Faraday's Second Law of Electrolysis is expressed mathematically as follows:

$$\frac{\text{Mass of element X deposited}}{\text{Mass of element Y deposited}} = \frac{\dfrac{\text{Relative atomic mass of X}}{\text{Valency of X}}}{\dfrac{\text{Relative atomic mass of Y}}{\text{Valency of Y}}}$$

Or,

$$\frac{\text{Mass of element X deposited}}{\text{Mass of element Y deposited}} = \frac{\text{Chemical equivalent of X}}{\text{Chemical equivalent of Y}}$$

In order to determine the mass of an element deposited during electrolysis, we can use the expression below:

$$m = \frac{MIt}{vF}$$

where m is the mass of element deposited, M is the atomic mass of the element, I is the current, t is the time in seconds, v is the valency of the element, and F is the Faraday constant.

Magnetic Field

Magnetic flux density

Magnetic flux density, B, is given by:

$$B = \frac{\text{Magnetic flux}}{\text{Area}}$$

$$B = \frac{\emptyset}{A}$$

The magnetic flux, $\emptyset$, is expressed as:

$$\emptyset = BA$$

If the plane is inclined at an angle, θ, to the magnetic field, then:

$$\emptyset = BA\sin\theta$$

Magnetic force on a moving charge

The force exerted on a moving charge in a magnetic field depends on the magnitude of the field, B, the speed, v, the magnitude of the charge, q, and the angle, θ, between the direction of the field and the velocity. It is given by:

$$F = Bvq\sin\theta$$

When a charge moves in a perpendicular direction to the field, then, θ = 90°C. This gives a maximum force as follows:

$$F = Bvq$$

Also,

$$F = Bvq = \frac{mv^2}{r} \text{ (for a charge moving in a circular path)}$$

Or, $Bq = \dfrac{mv}{r}$

Force on a current carrying conductor in a magnetic field

The magnitude of the force depends on the strength of the magnetic field, B, the current, I, the length of the wire (conductor), L. and the angle, θ, between the field and the wire. In a uniform field, the force is given by:

$$F = BILsin\theta$$

Wire/conductor moving in a magnetic field
If the wire is of length, L, speed, v, and the magnetic field, B, makes an angle, θ, with the wire, then the induced emf in the wire is given by:

$$E = BvLsin\theta$$

The Transformer
The transformer equation
If E_p, I_p, and E_s, I_s, denotes the emf and current in the primary and secondary coil respectively, and the power losses in the transformer are negligible, the power output is equal to the power input. This gives:

$$E_sI_s = E_pI_p$$

Or, $I_pN_p = I_sN_s$

Hence, $\dfrac{E_S}{E_P} = \dfrac{N_S}{N_P}$

where N_p and N_s, are the number of turns in the primary and secondary coils respectively.

The efficiency of a transformer is usually less than 100% due to the power loss in the transformer. The efficiency is expressed as:

$$\text{Efficiency, } E = \frac{Power\ \ output}{Power\ \ input} \times 100$$

Or, $E = \dfrac{E_S I_S}{E_P I_P} \times 100$

Alternating Current (A.C) Circuit

The effective (or root-mean-square (r.m.s)) value of the current in an a.c circuit is given by:

$$I_{r.m.s} = \frac{I_o}{\sqrt{2}}$$

where $I_{r.m.s}$ = root-mean-square (r.m.s) value of the current and I_o = the peak value of the current.

Similarly, the effective or r.m.s value of the voltage is:

$$V_{r.m.s} = \frac{V_o}{\sqrt{2}}$$

where V_o = peak value of the voltage

By applying ohm's law:

$$V_{rms} = RI_{rms} \quad \text{and} \quad V_o = I_oR$$

where R is the resistance of the circuit.

The instantaneous current in an a.c circuit is given by:

$$I = I_o\sin\omega t$$

where, I = Instantaneous current, I_o = peak current, and ωt = phase angle. Note that $\omega = 2\pi f$, where f is the frequency.

R.L.C Series Circuit

Inductive reactance

The inductive reactance, X_L in an a.c circuit is given by:

$$X_L = \omega L$$

Or $\quad X_L = 2\pi fL$

where ω = angular velocity, f = frequency and L = inductance of the inductor.

The voltage across the inductor is given by:

$$V_L = IX_L,$$

where I is the current in the circuit.

The energy stored in an inductor is given by:

$$E_L = \frac{1}{2}I^2L$$

Where I is the current across the inductor, and L is the inductance of the inductor.

Capacitive reactance

It is given by:

$$X_C = \frac{1}{\omega C} \quad \text{or} \quad X_C = \frac{1}{2\pi f C}$$

where C = capacitance of the capacitor.

The voltage across the capacitor is given by:

$$V_C = IX_C$$

Resistor and Inductor (R-L) circuit

In this type of circuit, the effective voltage in the circuit is given by:

$$V = \sqrt{V_R^2 + V_L^2}$$

where, V_R is the voltage across the resistor, and V_L is the voltage across the inductor.

Note that the voltage across the resistor is given by:

$$V_R = IR$$

For an R-L circuit, the impedance is given by:

$$Z = \sqrt{R^2 + X_L^2} \quad \text{or}$$

Or, $$Z = \frac{V}{I}$$

The phase angle for this type of circuit is given by:

$$\tan\emptyset = \frac{X_L}{R}$$

Or $\quad \tan\emptyset = \dfrac{V_L}{V_R}$

where $\emptyset$ is the phase angle

Resistor and Capacitor (R-C) circuit

In an R-C circuit, the effective voltage in the circuit is given by:

$$V = \sqrt{V_R^2 + V_C^2}$$

where, V_R is the voltage across the resistor, and V_C is the voltage across the capacitor.

For an R-C circuit, the impedance is given by:

$$Z = \sqrt{R^2 + X_C^2}$$

Or $\quad Z = \dfrac{V}{I}$

The phase angle for this type of circuit is given by:

$$\tan\emptyset = \dfrac{X_C}{R}$$

Or $\quad \tan\emptyset = \dfrac{V_C}{V_R}$

Inductor and Capacitor (L-C) circuit

In an L-C circuit, the effective voltage is given by:

$$V = V_L - V_C$$

The effective reactance/overall reactance or impedance in this type of circuit is given by:

$$Z = X_L - X_C$$

Resistor, Inductor and Capacitor (R-L-C) circuit

In an R-L-C circuit, the effective voltage in the circuit is given by:

$$V = \sqrt{V_R^2 + (V_L - V_C)^2}$$

where, V_R is the voltage across the resistor, V_L is the voltage across the inductor, and V_C is the voltage across the capacitor.

For an R-L-C circuit, the impedance is given by:

$$Z = \sqrt{R^2 + (X_L - X_C)^2}$$

Or, $\quad Z = \dfrac{V}{I}$

The phase angle for this type of circuit is given by:

$$\tan\emptyset = \frac{X_L - X_C}{R}$$

Or $\tan\emptyset = \dfrac{V_L - V_C}{V_R}$

Note that for all the circuit types, impedance is given by: $\quad Z = \dfrac{V}{I}$, thus, V = IZ

Resonance in an a.c circuit

At resonance, $X_L = X_C$, i.e. $\omega L = \dfrac{1}{\omega C}$

Or, $\quad 2\pi f_o L = \dfrac{1}{2\pi f_o C}$

where f_o is the resonance frequency.

Hence, the resonance frequency is given by:

$$f_o = \frac{1}{2\pi\sqrt{LC}}$$

Power in a.c circuit

Instantaneous power, P, consumed in an a.c circuit is given by:

47

$$P = IV$$

where I and V are the instantaneous values.

The average power consumed in an a.c circuit is given by:

$$P_{av} = \frac{I_0 V_0 CosQ}{2}$$

Or $P_{av} = IVcos\emptyset$, I and V are effective values (or r.m.s values), and $cos\emptyset$ = power factor. The power factor is given by:

$$Cos\emptyset = \frac{R}{Z}$$

The power lost or dissipated in the circuit is given by:

$$P = I^2 R$$

Radioactivity

Half life

If a radioactive substance undergoes decay after a certain time, then the fraction of the substance remaining is given by:

$$\text{Fraction remaining} = \frac{1}{2^n}$$

$$\text{where, n} = \frac{\text{Total time}}{\text{half life}}$$

The fraction of the substance that has decayed is given by:

$$\text{Fraction decayed} = 1 - \frac{1}{2^n}$$

Decay constant

The half life of a radioactive substance can be expressed as follows:

$$t_{half} = \frac{0.693}{\lambda}$$

where t_{half} = half-life, and λ = decay constant

For a radioactive substance, the initial count rate, N_o, and the final count rater N after a time t, are related by:

$$N = N_o e^{-\lambda t}$$

Or $\quad e^{\lambda t} = \dfrac{N_o}{N}$

Or, $\quad \ln\left(\dfrac{N_o}{N}\right) = \lambda t$

Structure of the atom

The Nucleus
The atomic mass/mass number of an atom is given by:

$$A = N + Z$$

where A = mass number, N = number of neutrons, and Z =number of protons

Nuclear fission
Binding energy for the nucleus of an atom is given by Albert Einstein equation as follows:

$$E = \Delta mc^2$$

where E is energy released (binding energy), Δm is the loss in mass and c is the velocity of light.

Atomic Energy Levels
The energy loss or gained when an electron jumps from one energy level to another is given by:

$$hf = \Delta E = E_h - E_L$$

Or, $\quad \dfrac{hc}{\lambda} = \Delta E$

where h = Planck's constant, ΔE = change in energy/energy loss or gained, c is the speed of light, λ is the wavelength, and E_h and E_L are energy of the higher orbit and lower orbit respectively. The lowest orbit which could be represented as E_o or E_1 is called the ground state energy.

The energy of any level is given by:

$$E_n = \frac{E_o}{n^2}$$

where n = 2, 3, 4, ... depending on the level.

The Franck-Hertz experiment

From Frank and Hertz experiment, the energy gained by a free electron in passing through a gas under the influence of an electric field is:

$$E = hf = \frac{hc}{\lambda}$$

where h = Planck's constant, f = frequency of the electron, λ = the wavelength of the emitted electron, and $c = 3 \times 10^8$ m/s = speed of light or emitted ray.

The kinetic energy of the electrons is given by:

$$\tfrac{1}{2}mv^2 = eV$$

where V = the accelerating potential on electron and $e = 1.6 \times 10^{-19}$C = charge of an electron. This kinetic energy is also be expressed as:

$$E = \tfrac{1}{2}mv^2 = eV = hf = \frac{hc}{\lambda}$$

Photoelectric effect

Photoelectricity shows that light of frequency, f, is made up of photon energy which is given by:

$$E = hf = \frac{hc}{\lambda}$$

Einstein's photoelectric equation

Work function, w, of a metal surface is given by:

$$w = hf_o = \frac{hc}{\lambda_o}$$

where f_o = threshold frequency, and λ_o = threshold wavelength. The maximum kinetic energy, E_m which an electron can attain as it leaves a metal surface is:

$$E_m = hf - w$$

Or, $\quad E_m = hf - hf_o$

This equation is called Einstein's photoelectric equation. It can also be expressed as:

$$\tfrac{1}{2}mv_m^2 = hf - hf_o$$

$$= \frac{hc}{\lambda} - \frac{hc}{\lambda_o}$$

where m is the mass of the electron, v_m is the maximum velocity of the emitted electrons.

The maximum kinetic energy of the electron is also related to the potential energy difference, V_s (stopping potential) of the electron. This is given by:

$$E_m = eV_s$$

Hence, the energy of an electron can be expressed in terms of potential difference as follows:

$$E = eV$$

where e = charge of an electron, and V is the applied potential difference.

Wave-Particle nature of matter

From Luois de Broglie's equation, the wavelength of a particle is given by:

$$\lambda = \frac{h}{mv}$$

where mv (m is mass and v is velocity) is the momentum of the particle and h is Plank's constant.

The uncertainty principle

The uncertainty in the momentum multiplied by the uncertainty in the position approximately equals the Plank's constant, h. Heisenberg showed that:

$$\Delta x \times \Delta p \geq h$$

$$\Delta x \times \Delta v \geq h$$

$$\Delta E \times \Delta t \geq h$$

where ΔE, Δt, Δx and Δv are the uncertainties in energy, time, position and velocity respectively.

PART 2: TABLES OF PHYSICAL CONSTANTS

Note: The information given in the tables in this book are values at room temperature, between 20°C and 25°C where applicable, unless otherwise stated.

Table of Fundamental Physical Constants

Name	Symbol	Value
Speed of light	c	2.99792458×10^8 m/s
Planck constant	h	$6.6260755 \times 10^{-34}$ Js
Planck constant	h	$4.1356692 \times 10^{-15}$ eVs
Planck h bar	$\hbar$	$1.0545727 \times 10^{-34}$ Js
Planck h bar	$\hbar$	6.582121×10^{-16} eVs
Gravitation constant	G	$6.67259 \times 10^{-11} m^3 Kg^{-1} s^{-2}$
Boltzmann constant	K	1.380658×10^{-23} J/K
Boltzmann constant	K	8.617385×10^5 eV/K
Molar gas constant	R	8.314510J/mol K
Avogadro's number	N_A	6.0221×10^{23} mol^{-1}
Charge of electron	e	$1.60217733 \times 10^{-19}$ C
Permeability of vacuum	μ_0	$4\pi \times 10^{-7}$ N/A^2
Permittivity of vacuum	ε_0	$8.854187817 \times 10^{-12}$ F/m
Coulomb constant	$K = \dfrac{1}{4\pi\varepsilon_0}$	8.987552×10^9 Nm^2/C^2
Faraday constant	F	96485.309C/mol
Mass of electron	m_e	$9.1093897 \times 10^{-31}$ Kg
Mass of electron	m_e	0.51099906MeV/c^2
Mass of proton	m_p	$1.6726231 \times 10^{-27}$ Kg
Mass of proton	m_p	938.27231MeV/c^2
Mass of neutron	m_n	$1.6749286 \times 10^{-27}$ Kg
Mass of neutron	m_n	939.56563MeV/c^2
Atomic mass unit	u	$1.6605402 \times 10^{-27}$ Kg
Atomic mass unit	u	931.49432MeV/c^2
Avogadro's number	N_A	6.0221367×10^{23}/mol
Stefan-Boltzmann constant	σ	$5.67051 \ 10^{-8}$ W/m^{2}K^4

Rydberg constant	R_∞	$10973731.534 \text{m}^{-1}$
Bohr magneton	μ_B	$9.2740154 \times 10^{-24} \text{J/T}$
Bohr magneton	μ_B	$5.788382 \times 10^{-5} \text{eV/T}$
Flux quantum	ϕ_o	$2.0678338 \times 10^{-15} \text{T m}^2$
Bohr radius	a_o	$0.529177249 \times 10^{-10} \text{m}$
Standard atmosphere	atm	101325Pa
Wien displacement constant	b	$2.897756 \times 10^{-3} \text{mK}$

Table of Relative Permittivity/Dielectric Constant of Materials

Material	Relative Permittivity (ε_r)/Dielectric Constant
Vacuum	1
Argon (68° F)	1.000513
Air (Dry) (68° F)	1.000536
Nitrogen (68° F)	1.000580
Carbon dioxide (68° F)	1.000921
Acetone (32° F)	1.0159
Acetylene (32° F)	1.0217
Air, Liquid (-191°C)	1.4
Butane (30° F)	1.4
Insulation of telephone cables	1.5
Chlorine (32° F)	2.0
Polytetrafluoroethylene (PTFE)	2
Wood, Dry	2 - 6
Isoprene (77° F)	2.1
Mineral Oil (80° F)	2.1
Paraffin Wax	2.1-2.5
Teflon, PTFE	2.1
Aluminum Fluoride	2.2
Paraffin oil	2.2
Petroleum	2.2
Polyethylene, XLPE	2.2-2.4

Polypropylene	2.2
Silicon oil	2.2 - 2.8
Transformer oil, mineral	2.2
Turpentine	2.2
Benzene (68° F)	2.3
Hydrocyanic acid (21°C)	2.3
Paper	2.3
R12 Dichlorodifluoromethane (70° F)	2.4
Casting compound	2.5
Ebonite	2.5 - 2.9
Mica	2.5 - 7
Paper, waxed	2.5
Silica Sand	2.5-3.5
Soft rubber	2.5
Transformer oil, vegetable	2.5
Vulcanized fibres	2.5
Carbon disulfide	2.6
Amber	2.8-2.9
Polyester Resin	2.8 - 4.5
Polyamide	2.8
Calcium	3.0
Olive oil	3
Polystyrene	3
Rubber	3.0
Salt	3.0 - 15.0
Mylar	3.1
Ice (-2°C)	3.2
Plexiglass	3.2
Aluminum Bromide (212° F)	3.4
Polymide	3.4
Bakelite	3.5-5.0
Shellac	3.5
Sulfur	3.5
Araldite	3.6
Epoxy Resin (Cast)	3.6

Glass	3.7 - 10
Silicon dioxide	3.9
Guttapercha	4
Nylon	4.0 - 5.0
Oil paper	4
Pressed board	4
Slate	4
Water	4 - 88
Acetic Acid (36° F)	4.1
Insulation of high voltage cables	4.2
Pyrex Glass	4.3 - 5.0
Concrete	4.5
Hard paper, laminated	4.5
Quartz	4.5
Caster oil	4.7
Chloroform (68° F)	4.8
Paper, impregnated	5
Porcelain	5.0-7.0
Aniline (212° F)	5.5
Diamond	5.5 - 10
Sodium chloride	5.6
Ethyl Acetate (77° F)	6.0
Steatite	6
Acetic Acid (68° F)	6.2
Neoprene	6.7
Granite	7 - 9
Aniline (68° F)	7.3
Aniline (32° F)	7.8
Body tissue	8
Marble	8
Phenolic resin	8
Sapphire	8.9 - 11.1
Alumina	9.3-11.5
Graphite	10 - 15
Silicon	11.0 - 12.0
Acetyl Chloride (68° F)	15.8

Alcohol, Industrial	16-31
Acetyl Bromide (68° F)	16.5
Ammonia (69° F)	16.5
Acetone (127° F)	17.7
Ammonia (40° F)	18.9
Acetone (77° F)	20.7
Ceramic, $MgNb_2O_6$	21
Acetaldehyde (41° F)	21.8
Ammonia (-30° F)	22.0
Acetyl Acetone (68° F)	23.1
Ethanol (25°C, 77° F)	24.3
Ethyl Alcohol (77° F)	24.3
Acetyle Acetone (68° F)	25.0
Ammonia (-74° F)	25
Ceramic, $ZnNb_2O_6$	25
Ceramic, $MgTa_2O_6$	28
Methanol (20°C))	33.6
Ethylene glycol	37
Ceramic, $ZnTa_2O_6$	38
Furfural	42
Glycerol (77° F)	42.5
Glycerin, Liquid	47-68
Hydrazine (20°C)	52
Hydrogen peroxide (25°C)	60
Hydrofluoric acid (0°C)	83.6
Formamide (20°C)	84
Sulfuric acid (20°C)	84
Titanium dioxide	86 - 173

Table of Specific Capacitance of Materials

Material	Specific Capacitance (F/g)
Tobacco	148
Onion	200
Banana stem	170

Rubber wood sawdust	138
Cow dung	124
Oil palm fruit	85
Corn stalk	140
Fructose corn syrup	168
Carbon black	95
Particulate from TiC	220
Carbon Cloth	200
Aerogel carbon	140
Particulate from SiC	175
Anhydrous RuO_2	150
Doped conducting polymers	450
Hydrous RuO_2	650

Table of Electrical Conductivity and Resistivity of Materials

Material	Resistivity, ρ, (Ωm) at 20 °C	Conductivity, σ, (S/m) at 20 °C
Silver	1.59×10^{-8}	6.30×10^{7}
Copper	1.68×10^{-8}	5.96×10^{7}
Annealed copper	1.72×10^{-8}	5.80×10^{7}
Gold	2.44×10^{-8}	4.10×10^{7}
Aluminum	2.82×10^{-8}	3.5×10^{7}
Calcium	3.36×10^{-8}	2.98×10^{7}
Tungsten	5.60×10^{-8}	1.79×10^{7}
Zinc	5.90×10^{-8}	1.69×10^{7}
Nickel	6.99×10^{-8}	1.43×10^{7}
Lithium	9.28×10^{-8}	1.08×10^{7}
Iron	1.0×10^{-7}	1.00×10^{7}
Platinum	1.06×10^{-7}	9.43×10^{6}
Tin	1.09×10^{-7}	9.17×10^{6}
Carbon steel	(10^{10})	1.43×10^{-7}
Lead	2.2×10^{-7}	4.55×10^{6}
Titanium	4.20×10^{-7}	2.38×10^{6}

Grain oriented electrical steel	4.60×10^{-7}	2.17×10^{6}
Manganin	4.82×10^{-7}	2.07×10^{6}
Constantan	4.9×10^{-7}	2.04×10^{6}
Stainless steel	6.9×10^{-7}	1.45×10^{6}
Mercury	9.8×10^{-7}	1.02×10^{6}
Nichrome	1.10×10^{-6}	9.09×10^{5}
GaAs	5×10^{-7} to 10×10^{-3}	5×10^{-8} to 10^{3}
Carbon (amorphous)	5×10^{-4} to 8×10^{-4}	1.25 to 2×10^{3}
Carbon (diamond)	1×10^{12}	$\sim 10^{-13}$
Germanium	4.6×10^{-1}	2.17
Sea water	2×10^{-1}	4.8
Drinking water	2×10^{1} to 2×10^{3}	5×10^{-4} to 5×10^{-2}
Silicon	6.40×10^{2}	1.56×10^{-3}
Wood (damp)	1×10^{3} to 4	10^{-4} to 10^{-3}
Deionized water	1.8×10^{5}	5.5×10^{-6}
Glass	10×10^{10} to 10×10^{14}	10^{-11} to 10^{-15}
Hard rubber	1×10^{13}	10^{-14}
Wood (oven dry)	1×10^{14} to 16	10^{-16} to 10^{-14}
Sulfur	1×10^{15}	10^{-16}
Air	1.3×10^{16} to 3.3×10^{16}	3×10^{-15} to 8×10^{-15}
Paraffin wax	1×10^{17}	10^{-18}
Fused quartz	7.5×10^{17}	1.3×10^{-18}
PET	10×10^{20}	10^{-21}
Teflon	10×10^{22} to 10×10^{24}	10^{-25} to 10^{-23}

Table of Temperature Coefficient of Materials

Material	Temperature coefficient (K^{-1})
Silver	3.80×10^{-3}
Copper	4.04×10^{-3}
Annealed copper	3.93×10^{-3}
Gold	3.40×10^{-3}
Aluminium	3.90×10^{-3}
Calcium	4.10×10^{-3}
Tungsten	4.50×10^{-3}
Zinc	3.70×10^{-3}

Cobalt	7.00×10^{-3}
Nickel	6.00×10^{-3}
Lithium	6.00×10^{-3}
Iron	5.00×10^{-3}
Platinum	3.92×10^{-3}
Tin	4.50×10^{-3}
Gallium	4.00×10^{-3}
Lead	3.90×10^{-3}
Titanium	3.80×10^{-3}
Manganin	0.002×10^{-3}
Constantan	0.008×10^{-3}
Stainless steel	0.94×10^{-3}
Mercury	0.90×10^{-3}
Nichrome	0.40×10^{-3}
Carbon (amorphous)	-0.50×10^{-3}
Germanium	-48.0×10^{-3}
Silicon	-75.0×10^{-3}

Table of Specific Heat Capacity of Materials

Material	Specific Heat capacity $(Jkg^{-1}K^{-1})$
Aluminium	887
Asphalt	915
Bone	440
Boron	1106
Brass	920
Brick	841
Cast Iron	554
Clay	878
Coal	1262
Cobalt	420
Concrete	879
Copper	385
Glass	792
Gold	130
Granite	774
Gypsum	1090

Helium	5192
Hydrogen	14300
Ice	2090
Iron	462
Lead	130
Limestone	806
Lithium	3580
Magnesium	1024
Marble	832
Mercury	126
Nitrogen	1040
Oak Wood	2380
Oxygen	919
Platinum	150
Plutonium	140
Quartzite	1100
Rubber	2005
Salt	881
Sand	780
Sandstone	740
Silicon	710
Silver	236
Soil	1810
Stainless Steel 316	468
Steam	2094
Sulfur	706
Thorium	118
Tin	226
Titanium	521
Tungsten	133
Uranium	115
Vandium	490
Water	4187
Zinc	389

Table of Density of Solids

Solid	Density (g/cm^3)
ABS - copolymer of acrylonitrile, butadiene and styrene	1.06
Acetals	1.42
Acrylic	1.19
Agate	2.5 - 2.7
Agate	2.6
Alabaster carbonate	2.7 - 2.8
Alabaster sulfate	2.3
Albite	2.6 - 2.65
Alum, lumpy	0.881
Alum, pulverized	0.752
Alumina (aluminium oxide)	3.95 - 4.1
Aluminum	2.7
Aluminum Bronze	7.7
Amber	1.06 - 1.1
Amphiboles	2.9 - 3.2
Andesite, solid	2.77
Anorthite	2.74 - 2.76
Antimony, cast	6.7
Arsenic	4.7
Artificial Wool	1.5
Asbestos	2.0 - 2.8
Asbestos, shredded	0.35
Asbestos, solid	2.45
Ashes	0.65
Asphalt, compacted	2.36
Asphalt, crushed	0.72
Bakelite	1.36
Baking powder	0.72
Balsa Wood	0.13
Barite, crushed	2.89
Barium	3.78
Bark, wood refuse	0.24
Barytes	4.5
Basalt	2.4 - 3.1
Bauxite, crushed	1.28

Beeswax	0.96
Beryl	2.7
Beryllia	3.0
Beryllium	1.85
Biotite	2.7 - 3.1
Bismuth	9.8
Boiler scale	2.5
Bone	1.7 - 2.0
Bone, pulverized	0.88
Borax, fine	0.85
Brasses	8.47 - 8.75
Brick	1.4 - 2.4
Brick, fire	2.3
Brick, hard	2
Brick, pressed	2.2
Brickwork in cement	1.8
Brickwork in mortar	1.6
Bronzes	8.74 - 8.89
Brown iron ore	5.1
Butter	0.86 - 0.87
Cadmium	8.64
Calamine	4.1 - 4.5
Calcium	1.55
Calcspar	2.6 - 2.8
Camphor	1
Caoutchouc	0.9 – 1
Carbon	3.51
Cardboard	0.7
Cast Iron	7.2
Celluloid	1.4
Cellulose acetate, moulded	1.22 - 1.34
Cellulose acetate, sheet	1.28 - 1.32
Cellulose nitrate, celluloid	1.35 - 1.4
Cellulose, cotton, wood pulp, regenerated	1.48 - 1.53
Cement, Portland	1.5
Cement, set	2.7 – 3
Cerium	6.77
Chalk	1.9 - 2.8

Charcoal, oak	0.6
Charcoal, pine	0.3 - 0.4
Chlorinated polyether	1.4
Chrom oxide	5.21
Chromium	7.1
Cinnabar	8.1
Clay	1.8 - 2.6
Coal, anthracite	1.4 - 1.8
Coal, bituminous	1.2 - 1.5
Cobalt	8.8
Cocoa, butter	0.9
Coke	1 - 1.7
Concrete, dense	2.0 - 2.4
Concrete, lightweight	0.45 - 1.0
Concrete, medium	1.3 - 1.7
Constantan	8.89
Copal	1 - 1.15
Copper	8.79
Cork	0.2 - 0.25
Cork, linoleum	0.55
Corundum	4.0
Cotton	0.08
CPVC - Chlorinated poly vinyl chloride	1.6
Diamond	3 - 3.5
Dolomite	2.8
Duralium	2.8
Earth, loose	1.2
Earth, rammed	1.6
Ebonite	1.15
Electron	1.8
Emery	4
Epidote	3.2 - 3.5
Epoxy cast resin	1.11 - 1.4
Epoxy glass fibre	1.5
Expanded polystyrene	0.015 - 0.03
Feldspar	2.6 - 2.8
Fire brick	1.8 - 2.2
Flint	2.6

Fluorite	3.2
Galena	7.3 - 7.6
Gallium	5.9
Gamboge	1.2
Garnet	3.2 - 4.3
Gas carbon	1.9
Gelatin	1.3
Germanium	5.32
Glass, common	2.4 - 2.8
Glass, flint	2.9 - 5.9
Glass, Pyrex	2.21
Glass-wool	0.025
Glue	1.3
Gneiss	2.69
Gold	19.29
Granite	2.6 - 2.8
Graphite	2.3 - 2.7
Gum arabic	1.3 - 1.4
Gypsum	2.3
Hardboard	1.0
HDPE - (PEH) - high density poly ethylene	0.96
Hematite	4.9 - 5.3
Hornblende	3
Ice	0.917
Iodine	4.95
Iridium	22.5
Iron, cast	7.0 - 7.4
Ivory	1.8 - 1.9
Kaolin	2.6
LDPE - low density poly ethylene	0.91
Lead	11.35
Lead Crystal	3.1
Leather, dry	0.86
Lime, slaked	1.35
Limestone	2.7 -2.8
Linoleum	1.2
Lithium	0.53
Magnesia	3.2 - 3.6

Magnesium	1.74
Magnetite	4.9 - 5.2
Malachite	3.7 - 4.1
Manganese	7.43
Marble	2.6 - 2.8
Meerschaum	1 - 1.3
Mica	2.6 - 3.2
Mineral wool quilt	0.05
Molybdenum	10.2
Muscovite	2.8 – 3
Nickel	8.9
Nylon 6	1.12 - 1.17
Nylon 6,6	1.13 - 1.15
Oak	0.72
Ochre	3.5
Opal	2.2
Osmium	22.48
Palladium	12.0
Paper	0.7 - 1.15
Paraffin	0.9
PBT - poly butylene terephthalate	1.35
PC - poly carbonate	1.2
Peat blocks	0.85
PET - poly ethylene terephthalate	1.35
Phenolic cast resin	1.24 - 1.32
Phosphorbronce	8.8
Phosphorus	1.82
Pinchbeck	8.65
Pit coal	1.35
Pitch	1.1
Plaster board	0.80
Platinum	21.5
Plywood	0.54
PMMA - poly methyl methacrylate	1.2
Polyacrylonitrile	1.16 - 1.18
Polyamides	1.15 - 1.25
POM - poly oxy methylene	1.4
Porcelain	2.3 - 2.5
Porphyry	2.6 - 2.9

Potassium	0.86
PP - poly propylene	0.91 - 0.94
PPO - poly penylene ether	1.1
Pressed wood, pulp board	0.19
PS - poly styrene	1.03
PTFE - poly tetra fluoro ethylene, Teflon	2.28 - 2.30
PU - poly urethane foam	0.03
PVC - poly vinyl chloride	1.39 - 1.42
PVDF - poly vinylidene fluoride	1.76
Pyrex	2.25
Pyrite	4.9 - 5.1
Quartz	2.65
Radium	5
Red lead	8.6 - 9.1
Red metal	8.8
Resin	1.07
Rhenium	21.4
Rhodium	12.3
Rock salt	2.2
Rock wool	0.22 - 0.39
Rosin	1.07
Rubber, foam	0.070
Rubber, hard	1.2
Rubber, pure gum	0.91 - 0.93
Rubber, soft commercial	1.1
Rubidium	1.52
Sand, dry	1.4 - 1.6
Sandstone	2.1 - 2.4
Sapphire	3.98
Selenium	4.4
Serpentine	2.5 - 2.65
Silica, fused transparent	2.2
Silica, translucent	2.1
Silicium carbide	3.16
Silicon	2.33
Silver	10.5
Slag	2 - 3.9
Slate	2.6 - 3.3

Snow	0.1
Soapstone	2.6 - 2.8
Sodium	0.98
Soil	2.05
Solder	8.7 - 9.4
Soot	1.6 - 1.7
Spermaceti	0.95
Starch	1.5
Steatite	2.6 - 2.7
Steel	7.82
Stone	2.3 - 2.8
Sugar	1.6
Sulfur, cryst.	2.0
Talc	2.7 - 2.8
Tallow, beef	0.95
Tallow, mutton	0.95
Tantalum	16.6
Tar	1.05
Teflon	2.20
Tellurium	6.25
Thoria	4.16
Thorium	11.7
Tin	7.31
Titanium	4.5
Topaz	3.5 - 3.6
Tourmaline	3 - 3.2
Tungsten	19.2
Tungsten carbide	14.0 - 15.0
Uranium	19.1
Urethane foam (Urea formaldehyde foam)	0.08
Vanadium	6.1
Vermiculite	0.12
Wax, sealing	1.8
White metal	7.5 – 10
Wood wool slab	0.5 - 0.8
Zinc	7.14

Table of Density of Liquids

Liquid	Temperature (°C)	Density (Kg/m³)	Liquid	Temperature (°C)	Density (kg/m³)
Acetic Acid	25	1049	Heptane	25	679.5
Acetone	25	784.6	Hexane	25	654.8
Acetonitrile	20	782	Hexanol	25	811
Alcohol, ethyl (ethanol)	25	785.1	Hexene	25	671
Alcohol, methyl (methanol)	25	786.5	Hydrazine	25	795
Alcohol, propyl	25	800	Ionene	25	932
Ammonia (aqua)	25	823.5	Isobutyl Alcohol	20	802
Aniline	25	1019	Iso-Octane	20	692
Automobile oils	15	880 - 940	Isopropyl Alcohol	20	785
Beer (varies)	10	1010	Isopropyl Myristate	20	853
Benzene	25	873.8	Kerosene	60°F	820.1
Benzil	15	1230	Linolenic Acid	25	897
Brine	15	1230	Linseed oil	25	929.1
Bromine	25	3120	Machine oil	20	910
Butyric Acid	20	959	Mercury		13590
Butane	25	599	Methane	-164	465
n-Butyl Acetate	20	880	Methanol	20	791
n-Butyl Alcohol	20	810	Methyl Isoamyl Ketone	20	888
n-Butylhloride	20	886	Methyl Isobutyl Ketone	20	801
Caproic acid	25	921	Methyl n-Propyl Ketone	20	808
Carbolic acid	15	956	Methyl t-Butyl	20	741

			Ether		
Carbon disulfide	25	1261	N-Methylpyrrolidone	20	1030
Carbon tetrachloride	25	1584	Methyl Ethyl Ketone	20	805
Carene	25	857	Milk	15	1020 - 1050
Castor oil	25	956.1	Naphtha	15	665
Chloride	25	1560	Naphtha, wood	25	960
Chlorobenzene	20	1106	Napthalene	25	820
Chloroform	20	1489	Nitric acid	0	1560
Chloroform	25	1465	Ocimene	25	798
Citric acid, 50% aqueous solution	15	1220	Octane	15	698.6
Coconut oil	15	924	Oil of resin	20	940
Cotton seed oil	15	926	Oil of turpentine	20	870
Cresol	25	1024	Oil, lubricating	20	900
Creosote	15	1067	Olive oil	20	800 - 920
Crude oil, 48o API	60°F	790	Oxygen (liquid)	-183	1140
Crude oil, 40o API	60°F	825	Paraffin		800
Crude oil, 35.6o API	60°F	847	Palmitic Acid	25	851
Crude oil, 32.6o API	60°F	862	Pentane	20	626
Crude oil,alifornia	60°F	915	Pentane	25	625
Crude oil, Mexican	60°F	973	Perchlor ethylene	20	1620
Crude oil, Texas	60°F	873	Petroleum Ether	20	640
Cumene	25	860	Petrol, natural	60°F	711

Cyclohexane	20	779	Petrol, Vehicle	60°F	737
Cyclopentane	20	745	Phenol	25	1072
Decane	25	726.3	Phosgene	0	1378
Diesel fuel oil 20 to 60	15	820 - 950	Phytadiene	25	823
Diethyl ether	20	714	Pinene	25	857
o-Dichlorobenzene	20	1306	Propane	-40	493.5
Dichloromethane	20	1326	Propane, R-290	25	494
Diethylene glycol	15	1120	Propanol	25	804
Dichloromethane	20	1326	Propylenearbonate	20	1201
Dimethyl Acetamide	20	942	Propylene	25	514.4
N,N - Dimethylformamide	20	949	Propylene glycol	25	965.3
Dimethyl Sulfoxide	20	1100	Pyridine	25	979
Dodecane	25	754.6	Pyrrole	25	966
Ethane	-89	570	Rape seed oil	20	920
Ether	25	713.5	Resorcinol	25	1269
Ethylamine	16	681	Rosin oil	15	980
Ethyl Acetate	20	901	Sea water	25	1025
Ethyl Alcohol (Ethanol, pure alcohol, grain alcohol or drinking alcohol)	20	789	Silane	25	718
Ethyl Ether	20	713	Silicone oil	25	965 - 980
Ethylene Dichloride	20	1253	Sodium Hydroxide (caustic soda)	15	1250

Ethylene glycol	25	1097	Sorbaldehyde	25	895
Freon	25	1476	Soya bean oil	15	891
Fluorine refrigerant R-22	25	1194	Sulfuric Acid 95% onc.	20	1839
Formaldehyde	45	812	Sulfurus acid	-20	1490
Formic acid 10%oncentration	20	1025	Sugar solution 68 brix	15	1338
Formic acid 80%oncentration	20	1221	Sunflower oil	20	920
Freon - 11	21	1490	Styrene	25	903
Freon - 21	21	1370	Terpinene	25	847
Fuel oil	60°F	890	Tetrahydrofuran	20	888
Furan	25	1416	Toluene	20	867
Furforol	25	1155	Trichlor ethylene	20	1470
Gasoline, natural	60°F	711	Triethylamine	20	728
Gasoline, Vehicle	60°F	737	Trifluoroacetic Acid	20	1489
Gas oils	60°F	890	Turpentine	25	868.2
Glucose	60°F	1350 – 1440	Water, heavy	11.6	1105
Glycerine	25	1259	Water - pure	4	1000
Glycerol	25	1126	Water - sea	77°F	1022
Heating oil	20	920	Whale oil	15	925

Table of Thermal Conductivity of Elements

Metal	Thermal Conductivity (W/cmK)
Radon	0.0000364
Xenon	0.0000569

Chlorine	0.000089
Krypton	0.0000949
Argon	0.0001772
Nitrogen	0.0002598
Oxygen	0.0002674
Fluorine	0.000279
Neon	0.000493
Bromine	0.00122
Helium	0.00152
Hydrogen	0.001815
Phosphorus	0.00235
Sulfur	0.00269
Iodine	0.00449
Astatine	0.017
Selenium	0.0204
Tellurium	0.0235
Neptunium	0.063
Plutonium	0.0674
Manganese	0.0782
Bismuth	0.0787
Mercury	0.0834
Lawrencium	0.1
Einsteinium	0.1
Berkelium	0.1
Californium	0.1
Fermium	0.1
Curium	0.1
Nobelium	0.1
Americium	0.1
Mendelevium	0.1
Gadolinium	0.106
Dysprosium	0.107
Terbium	0.111
Cerium	0.114
Actinium	0.12
Praseodymium	0.125
Samarium	0.133
Lanthanum	0.135
Europium	0.139

Erbium	0.143
Francium	0.15
Scandium	0.158
Holmium	0.162
Lutetium	0.164
Neodymium	0.165
Thulium	0.168
Yttrium	0.172
Promethium	0.179
Barium	0.184
Radium	0.186
Polonium	0.2
Titanium	0.219
Zirconium	0.227
Hafnium	0.23
Rutherfordium	0.23
Antimony	0.243
Boron	0.274
Uranium	0.276
Vanadium	0.307
Ytterbium	0.349
Lead	0.353
Strontium	0.353
Cesium	0.359
Gallium	0.406
Thallium	0.461
Protactinium	0.47
Rhenium	0.479
Arsenic	0.502
Technetium	0.506
Niobium	0.537
Thorium	0.54
Tantalum	0.575
Dubnium	0.58
Rubidium	0.582
Germanium	0.599
Tin	0.666
Platinum	0.716
Palladium	0.718

Iron	0.802
Indium	0.816
Lithium	0.847
Osmium	0.876
Nickel	0.907
Chromium	0.937
Cadmium	0.968
Cobalt	1
Potassium	1.024
Zinc	1.16
Ruthenium	1.17
Carbon	1.29
Molybdenum	1.38
Sodium	1.41
Iridium	1.47
Silicon	1.48
Rhodium	1.5
Magnesium	1.56
Tungsten	1.74
Calcium	2.01
Beryllium	2.01
Aluminum	2.37
Gold	3.17
Copper	4.01
Silver	4.29

Note that these values in W/cmK can be converted to values in W/mK by simply multiplying by 100. The values are at room temperature.

Table of Thermal Conductivity of other materials

Material	Thermal Conductivity (W/mK)
Acetals	0.23
Acetone	0.16
Acetylene (gas)	0.018
Acrylic	0.2
Air, atmosphere (gas)	0.0262

Air, elevation 10000 m	0.020
Agate	10.9
Alcohol	0.17
Alumina	36
Aluminum Brass	121
Aluminum Oxide	30
Ammonia (gas)	0.0249
Apple (85.6% moisture)	0.39
Asbestos-cement board	0.744
Asbestos-cement sheets	0.166
Asbestos-cement	2.07
Asbestos, loosely packed	0.15
Asbestos mill board	0.14
Asphalt	0.75
Balsa wood	0.048
Bitumen	0.17
Bitumen/felt layers	0.5
Beef, lean (78.9 % moisture)	0.43 - 0.48
Benzene	0.16
Bitumen	0.17
Blast furnace gas (gas)	0.02
Boiler scale	1.2 - 3.5
Breeze block	0.10 - 0.20
Brick dense	1.31
Brick, fire	0.47
Brick, insulating	0.15
Brickwork, common (Building Brick)	0.6 -1.0
Brickwork, dense	1.6
Brown iron ore	0.58
Butter (15% moisture content)	0.20
Calcium silicate	0.05
Carbon dioxide (gas)	0.0146
Carbon monoxide	0.0232
Cellulose, cotton, wood pulp and regenerated	0.23
Cellulose acetate, molded, sheet	0.17 - 0.33
Cellulose nitrate, celluloid	0.12 - 0.21
Cement, Portland	0.29
Cement, mortar	1.73

Chalk	0.09
Charcoal	0.084
Chlorinated poly-ether	0.13
Chrome Nickel Steel	16.3
Chrom-oxide	0.42
Clay, dry to moist	0.15 - 1.8
Clay, saturated	0.6 - 2.5
Coal	0.2
Cod (83% moisture content)	0.54
Coke	0.184
Concrete, lightweight	0.1 - 0.3
Concrete, medium	0.4 - 0.7
Concrete, dense	1.0 - 1.8
Concrete, stone	1.7
Corian (ceramic filled)	1.06
Cork board	0.043
Cork, re-granulated	0.044
Cork	0.07
Cotton	0.04
Cotton wool	0.029
Cotton Wool insulation	0.029
Cupronickel 30%	30
Diamond	1000
Diatomaceous earth (Sil-o-cel)	0.06
Diatomite	0.12
Earth, dry	1.5
Ebonite	0.17
Emery	11.6
Engine Oil	0.15
Ethane (gas)	0.018
Ether	0.14
Ethylene (gas)	0.017
Epoxy	0.35
Ethylene glycol	0.25
Feathers	0.034
Felt insulation	0.04
Fiberglass	0.04
Fiber insulating board	0.048
Fiber hardboard	0.2

Fire-clay brick 500°C	1.4
Foam glass	0.045
Dichlorodifluoromethane R-12 (gas)	0.007
Dichlorodifluoromethane R-12 (liquid)	0.09
Gasoline	0.15
Glass	1.05
Glass, Pearls, dry	0.18
Glass, Pearls, saturated	0.76
Glass, window	0.96
Glass, wool Insulation	0.04
Glycerol	0.28
Granite	1.7 - 4.0
Graphite	168
Gravel	0.7
Ground or soil, very moist area	1.4
Ground or soil, moist area	1.0
Ground or soil, dry area	0.5
Ground or soil, very dry area	0.33
Gypsum board	0.17
Hairfelt	0.05
Hardboard high density	0.15
Hardwoods (oak, maple..)	0.16
Hastelloy C	12
Honey (12.6% moisture content)	0.5
Hydrochloric acid (gas)	0.013
Hydrogen sulfide (gas)	0.013
Ice (0°C, 32°F)	2.18
Inconel	15
Ingot iron	47 - 58
Insulation materials	0.035 - 0.16
Iron-oxide	0.58
Kapok insulation	0.034
Kerosene	0.15
Leather, dry	0.14
Limestone	1.26 - 1.33
Magnesia insulation (85%)	0.07
Magnesite	4.15
Magnesium alloy	70 - 145

Marble	2.08 - 2.94
Methane (gas)	0.030
Methanol	0.21
Mica	0.71
Milk	0.53
Mineral wool insulation materials, wool blankets ..	0.04
Neoprene	0.05
Nitric oxide (gas)	0.0238
Nitrous oxide (gas)	0.0151
Nylon 6, Nylon 6/6	0.25
Oil, machine lubricating SAE 50	0.15
Olive oil	0.17
Paper	0.05
Paraffin Wax	0.25
Peat	0.08
Perlite, atmospheric pressure	0.031
Perlite, vacuum	0.00137
Phenolic cast resins	0.15
Phenol-formaldehyde moulding compounds	0.13 - 0.25
Phosphorbronze	110
Pinchbeck	159
Pitch	0.13
Pit coal	0.24
Plaster light	0.2
Plaster, metal lath	0.47
Plaster, sand	0.71
Plaster, wood lath	0.28
Plasticine	0.65 - 0.8
Plastics, foamed (insulation materials)	0.03
Plywood	0.13
Polycarbonate	0.19
Polyester	0.05
Polyethylene low density, PEL	0.33
Polyethylene high density, PEH	0.42 - 0.51
Polyisoprene natural rubber	0.13
Polyisoprene hard rubber	0.16

Polymethylmethacrylate	0.17 - 0.25
Polypropylene, PP	0.1 - 0.22
Polystyrene, expanded	0.03
Polystyrol	0.043
Polyurethane foam	0.03
Porcelain	1.5
Potato, raw flesh	0.55
Propane (gas)	0.015
Polytetrafluoroethylene (PTFE)	0.25
Polyvinylchloride, PVC	0.19
Pyrex glass	1.005
Quartz mineral	3
Rock, solid	2 - 7
Rock, porous volcanic (Tuff)	0.5 - 2.5
Rock Wool insulation	0.045
Rosin	0.32
Rubber, cellular	0.045
Rubber, natural	0.13
Salmon (73% moisture content)	0.50
Sand, dry	0.15 - 0.25
Sand, moist	0.25 - 2
Sand, saturated	2 - 4
Sandstone	1.7
Sawdust	0.08
Sheep wool	0.039
Silica aerogel	0.02
Silicon cast resin	0.15 - 0.32
Silicon carbide	120
Silicon oil	0.1
Slag wool	0.042
Slate	2.01
Snow (temp < 0°C)	0.05 - 0.25
Softwoods (fir, pine ..)	0.12
Soil, clay	1.1
Soil, with organic matter	0.15 - 2
Soil, saturated	0.6 - 4
Solder 50-50	50
Soot	0.07
Steam, saturated	0.0184

Steam, low pressure	0.0188
Steatite	2
Straw slab insulation, compressed	0.09
Styrofoam	0.033
Sulfur dioxide (gas)	0.0086
Sulfur, crystal	0.2
Sugars	0.087 - 0.22
Tar	0.19
Timber, alder	0.17
Timber, ash	0.16
Timber, birch	0.14
Timber, larch	0.12
Timber, maple	0.16
Timber, oak	0.17
Timber, pitchpine	0.14
Timber, pockwood	0.19
Timber, red beech	0.14
Timber, red pine	0.15
Timber, white pine	0.15
Timber, walnut	0.15
Urethane foam	0.021
Vacuum	**0**
Vermiculite granules	0.065
Vinyl ester	0.25
Water	0.606
Water, vapor (steam)	
Wheat flour	0.45
White metal	35 - 70
Wood across the grain, white pine	0.12
Wood across the grain, balsa	0.055
Wood across the grain, yellow pine, timber	0.147
Wood, oak	0.17
Wool, felt	0.07
Wood wool, slab	0.1 - 0.15

Table of Melting Points of Elements

Element	Melting Point (°C)	Melting Point (K)	Melting Point (°F)
Helium	-272.05	1.1	-458
Hydrogen	-258.975	14.175	-434
Neon	-248.447	24.703	-415.205
Oxygen	-222.65	50.5	-368.77
Fluorine	-219.52	53.63	-363.14
Nitrogen	-209.86	63.29	-345.75
Argon	-189.19	83.96	-308.54
Krypton	-157.22	115.93	-251
Xenon	-111.7	161.45	-169.1
Chlorine	-100.84	172.31	-149.51
Radon	-71	202	-96
Mercury	-38.72	234.43	-37.7
Bromine	-7.1	266.05	19.2
Francium	27	300	81
Cesium	28.55	301.7	83.39
Gallium	29.9	303.05	85.8
Rubidium	39.64	312.79	103.35
Phosphorus	44.3	317.45	111.7
Potassium	63.35	336.5	146.03
Sodium	98	371	208
Iodine	113.5	386.65	236.3
Sulfur	115.36	388.51	239.65
Indium	156.76	429.91	314.17
Lithium	180.7	453.85	357.3
Selenium	221	494	430
Tin	232.06	505.21	449.71
Polonium	254	527	489
Bismuth	271.52	544.67	520.74
Astatine	302	575	576
Thallium	304	577	579
Cadmium	321.18	594.33	610.12
Lead	327.6	600.75	621.7
Zinc	419.73	692.88	787.51
Tellurium	449.65	722.8	841.37
Antimony	630.9	904.05	1167.6

Neptunium	640	913	1184
Plutonium	640	913	1184
Magnesium	649	922	1200
Aluminum	660.25	933.4	1220.45
Radium	700	973	1292
Barium	729	1002	1344
Strontium	769	1042	1416
Cerium	798	1071	1468
Arsenic	808	1081	1486
Europium	822	1095	1512
Ytterbium	824	1097	1515
Calcium	839	1112	1542
Einsteinium	860	1133	1580
Californium	900	1173	1652
Lanthanum	920	1193	1688
Praseodymium	931	1204	1708
Promethium	931	1204	1708
Germanium	937.4	1210.55	1719.3
Silver	961	1234	1762
Berkelium	986	1259	1807
Americium	994	1267	1821
Neodymium	1016	1289	1861
Actinium	1050	1323	1922
Gold	1064.58	1337.73	1948.24
Curium	1067	1340	1953
Samarium	1072	1345	1962
Copper	1084.6	1357.75	1984.3
Uranium	1132	1405	2070
Manganese	1244	1517	2271
Beryllium	1278	1551	2332
Gadolinium	1312	1585	2394
Terbium	1357	1630	2475
Silicon	1410	1683	2570
Dysprosium	1412	1685	2574
Nickel	1453	1726	2647
Holmium	1470	1743	2678
Cobalt	1495	1768	2723
Erbium	1522	1795	2772
Yttrium	1526	1799	2779

Iron	1535	1808	2795
Scandium	1539	1812	2802
Thulium	1545	1818	2813
Palladium	1552	1825	2826
Titanium	1660	1933	3020
Lutetium	1663	1936	3025
Thorium	1755	2028	3191
Platinum	1772	2045	3222
Protactinium	1600	2113	2912
Zirconium	1852	2125	3366
Chromium	1857	2130	3375
Vanadium	1902	2175	3456
Rhodium	1966	2239	3571
Rutherfordium	2127	2400	3860.6
Technetium	2200	2473	3992
Hafnium	2227	2500	4041
Ruthenium	2250	2523	4082
Boron	2300	2573	4172
Iridium	2443	2716	4429
Niobium	2468	2741	4474
Molybdenum	2617	2890	4743
Tantalum	2996	3269	5425
Osmium	3027	3300	5481
Rhenium	3180	3453	5756
Tungsten	3407	3680	6165
Carbon	3500	3773	6332

Table of Melting Points of Alloys and Chemicals

Alloy/Chemical	Melting Point (°C)	Melting Point (°F)
Aluminum-Cadmium Alloy	1377	2511
Aluminum-Calcium Alloy	545	1013
Aluminum-Cerium Alloy	655	1211
Aluminum-Copper Alloy	548	1018

Aluminum-Germanium Alloy	427	801
Aluminum-Gold Alloy	569	1056
Aluminum-Indium Alloy	637	1179
Aluminum-Iron Alloy	1153	2107
Aluminum-Magnesium Alloy	600	1110
Aluminum-Nickel Alloy	1385	2525
Aluminum-Platinum Alloy	1260	2300
Aluminum-Scandium Alloy	655	1211
Aluminum-Silicon Alloy	577	1071
Aluminum-Zinc Alloy	382	720
Amalgam	178-278	352.4-532.4
Arsenic-Antimony Alloy	605	1121
Arsenic-Cobalt Alloy	916	1681
Arsenic-Copper Alloy	685	1265
Arsenic-Indium Alloy	942	1728
Arsenic-Iron Alloy	1103	2017
Arsenic-Manganese Alloy	870	1598
Arsenic-Nickel Alloy	967	1770
Arsenic-Tin Alloy	579	1074
Arsenic-Zinc Alloy	1015	1859
Babbitt Metal	433-466	811.4-870.8
Beryllium-Copper Alloy	865 - 955	1587 - 1750
Brass	930	1710
Brass, Admiralty	900 - 940	1650 - 1720
Brass, Red	990 - 1025	1810 - 1880
Brass, Yellow	905 - 932	1660 - 1710
Bronze, Aluminum	1027 - 1038	1881 - 1900
Bronze, Manganese	865 - 890	1590 - 1630
Copper-Nickel Alloy	1060-1240	1940-2264
Field's Metal	62	144

Gold-Antimony Alloy	360	680
Gold-Bismuth Alloy	241	466
Gold-Cadmium Alloy	500	932
Gold-Cerium Alloy	520	968
Gold-Germanium Alloy	356	673
Gold-Lanthanum Alloy	561	1042
Gold-Lead Alloy	215	419
Gold-Magnesium Alloy	575	1067
Gold-Manganese Alloy	960	1760
Gold-Silicon Alloy	363	685
Gold-Sodium Alloy	876	1609
Gold-Tellurium Alloy	416	781
Gold-Thallium Alloy	131	268
Gold-Tin Alloy	278	532
Hastelloy C-276	1323-1371	2415-2500
Incoloy	1390 - 1425	2540 - 2600
Inconel	1390 - 1425	2540 - 2600
Invar	1427	2600
Iron, Cast	1204	2200
Iron, Cast (Gray)	1175 - 1290	2150 - 2360
Iron, Ductile	1,150 - 1,200	2,100 – 2,190
Iron, Wrought	1482	2700
Iron-Antimony Alloy	748	1378
Iron-Gadolinium Alloy	850	1562
Iron-Molybdenum Alloy	1452	2646
Iron-Niobium Alloy	1370	2498
Iron-Silicon Alloy	1202	2196
Iron-Tin Alloy	1127	2061
Iron-Yttrium Alloy	900	1652
Iron-Zirconium Alloy	1327	2421
Kovar	1449	2640
Lead-Antimony Alloy	247	477
Lead-Platinum Alloy	290	554
Lead-Praseodymium Alloy	1042	1908
Lead-Tellurium Alloy	924	1695
Lead-Tin Alloy	187	369
Lead-Titanium Alloy	725	1337

Magnesium AZ31B	~650	~1200
Magnesium-Antimony Alloy	961	1761.8
Magnesium-Nickel Alloy	507	945
Magnesium-Praseodymium Alloy	585	1085
Magnesium-Silicon Alloy	950	1742
Magnesium-Strontium Alloy	426	799
Magnesium-Zinc Alloy	342	648
Molybdenum-Nickel Alloy	1317	2403
Molybdenum-Niobium Alloy	2297	4167
Molybdenum-Osmium Alloy	2377	4311
Molybdenum-Rhenium Alloy	2507	4545
Molybdenum-Ruthenium Alloy	1927	3501
Molybdenum-Silicon Alloy	2077	3771
Monel	1300 - 1350	2370 - 2460
Nickel-Antimony Alloy	1102	2016
Nickel-Tin Alloy	1130	2066
Nickel-Titanium Alloy	1117	2043
Nickel-Tungsten Alloy	1500	2732
Nickel-Vanadium Alloy	1200	2192
Nickel-Zinc Alloy	875	1607
Nitinol	1300	2370
Pewter	240	464
Rose's Metal	98	208
Silver-Aluminum Alloy	562	1044
Silver-Antimony Alloy	485	905
Silver-Arsenic Allo	540	1004
Silver-Calcium Alloy	547	1017
Silver-Cerium Alloy	525	977

Silver-Copper Alloy	777	1431
Silver-Germanium Alloy	651	1204
Silver-Lanthanum Alloy	518	964
Silver-Lead Alloy	304	579
Silver-Lithium Alloy	145	293
Silver-Magnesium Alloy	472	882
Silver-Palladium Alloy	651	1204
Silver-Silicon Alloy	837	1539
Silver-Strontium Alloy	436	817
Silver-Tellurium Alloy	350	662
Silver-Zirconium Alloy	827	1521
Steel, Carbon	1425 - 1540	2600 - 2800
Steel, Maraging	1413	2575
Steel, Stainless	1510	2750
Stellite	1180-1415	2156 – 2579
Sterling Silver	893	1640
Titanium-Aluminum-Vanadium (Ti-6Al-4V)	1604 - 1660	2920 - 3020
Wood's Metal	70	158
2-propanol	-89.5	-129.1
Acetic acid	16.77	62.6
Acetone	-94	-137.2
Agar	85	185
Alcohol, ethyl (ethanol)	-114.38	-173.9
Alcohol, methyl (methanol)	-97.5	-143.5
Ammonium	-77.65	-107.77
Ammonium Nitrate	169.7	337.46
Beeswax	64	140
Benzene	5.72	42.3
Boric Acid	170.88	339.6
Canola Oil	-10	14
Carbon Dioxide	-56.6	-69.9
Carbon Monoxide	-120.6	-185.08
Carbonic Acid	210	410

Chloroform	-63.4	-82.12
Citric Acid	153	307.4
Dextrose	146	294.8
Ethlyne	-169.22	-272.6
Ethylene Dichloride	-35.5	-31.9
Ethylene Glycol	-12.8	8.96
Fructose	103	217.4
Glucose	146	294.8
Glycerine	17.77	64
Hexane	-95	-139
Hydrochloric Acid	−26	-14.8
Hydrofluoric Acid	-83.55	-118.4
Hydrogen Peroxide	-0.42	31.23
Isopropyl Alcohol	-89	-128.2
Kerosene	24-25	75.2-77
Lauric Acid	44	111.2
Methanol	-97.61	-143.7
Nitric Acid	-42	-43.6
Palmitic Acid	63	145.4
Paraffin	65.6	150
Phosphoric Acid	42.3	108.2
Polystyrene	240	464
Polyvinyl Chloride	100-260	212-500
Propane	-188	-306.4
Propylene	-185.11	-301.2
Propylene glycol	-60	-76
Silica (silicon dioxide)	1710	3110
Sodium Chloride	801	1474
Sodium Hydroxide	323	613.4
Sodium Hypochlorite	18	64.4
Stearic Acid	71.2	160.2
Sucrose	186	366.8
Sulfuric Acid	10.31	50.558
Toluene	-95	-139
Water, Fresh	0	32
Water, Sea	-2.38	27.7

Table of Boiling Point of Elements

Element	Boiling Point (°C)	Boiling Point (K)	Boiling Point (°F)
Helium	-268.785	4.365	-451.813
Hydrogen	-252.732	20.418	-422.918
Neon	-245.904	27.246	-410.6
Nitrogen	-195.65	77.5	-320.17
Fluorine	-188.05	85.1	-306.49
Argon	-185.7	87.45	-302.3
Oxygen	-182.82	90.33	-297.08
Krypton	-153.2	119.95	-243.8
Xenon	-107.97	165.18	-162
Radon	-62	211	-80
Chlorine	-33.9	239.25	-29
Bromine	59.25	332.4	138.65
Iodine	185.4	458.55	365.7
Phosphorus	280	553	536
Astatine	337	610	639
Mercury	357	630	675
Sulfur	444.75	717.9	832.55
Arsenic	603	876	1117
Cesium	671	944	1240
Francium	677	950	1251
Selenium	685	958	1265
Rubidium	688	961	1270
Potassium	759	1032	1398
Cadmium	765	1038	1409
Sodium	883	1156	1621
Zinc	907	1180	1665
Polonium	962	1235	1764
Tellurium	988	1261	1810
Magnesium	1090	1363	1994
Ytterbium	1194	1467	2181
Lithium	1342	1615.15	2448
Strontium	1384	1657	2523
Thallium	1473	1746	2683
Calcium	1484	1757	2703
Radium	1536	1809	2797

Bismuth	1564	1837	2847
Antimony	1587	1860	2889
Europium	1597	1870	2907
Lead	1740	2013	3164
Samarium	1791	2064	3256
Barium	1898	2171	3448
Thulium	1947	2220	3537
Manganese	1962	2235	3564
Indium	2073	2346	3763
Silver	2163	2436	3925
Tin	2270	2543	4118
Silicon	2355	2628	4271
Gallium	2403	2676	4357
Aluminum	2467	2740	4473
Dysprosium	2562	2835	4644
Copper	2567	2840	4653
Americium	2607	2880	4725
Chromium	2672	2945	4842
Holmium	2695	2968	4883
Nickel	2732	3005	4950
Iron	2750	3023	4982
Gold	2807	3080	5085
Germanium	2830	3103	5126
Scandium	2831	3104	5128
Erbium	2863	3136	5185
Cobalt	2870	3143	5198
Palladium	2964	3237	5367
Beryllium	2970	3243	5378
Terbium	3023	3296	5473
Neodymium	3068	3341	5554
Curium	3110	3383	5630
Actinium	3200	3473	5792
Plutonium	3230	3503	5846
Gadolinium	3266	3539	5911
Titanium	3287	3560	5949
Yttrium	3338	3611	6040
Lutetium	3395	3668	6143
Vanadium	3409	3682	6168
Cerium	3426	3699	6199

Lanthanum	3457	3730	6255
Promethium	3512	3785	6354
Praseodymium	3512	3785	6354
Rhodium	3727	4000	6741
Platinum	3827	4100	6921
Ruthenium	3900	4173	7052
Neptunium	3902	4175	7056
Boron	4002	4275	7236
Protactinium	4027	4300	7281
Uranium	4134	4407	7473
Zirconium	4377	4650	7911
Iridium	4428	4701	8002
Hafnium	4603	4876	8317
Molybdenum	4612	4885	8334
Niobium	4744	5017	8571
Thorium	4788	5061	8650
Carbon	4827	5100	8721
Technetium	4877	5150	8811
Osmium	5012	5285	9054
Tantalum	5425	5698	9797
Rutherfordium	5527	5800	9980.6
Rhenium	5627	5900	10161
Tungsten	5655	5928	10211

Table of Boiling Point of Common Chemicals

Element	Boiling Point (^{o}C)	Boiling Point (^{o}F)
Acetaldehyde	20.8	69
Acetic Acid Anhydride	139	282
Acetone	50.5	133
Acetylene	-84	-119
Alcohol - allyl	97.2	207
Alcohol - butyl-n	117	243
Alcohol - ethyl (grain, ethanol)	79	172.4
Alcohol - methyl (wood, methanol)	64.7	151
Alcohol - propyl	97.5	207

Ammonia	-35.5	-28.1
Aniline	184.4	363
Benzene (Benzol)	80.4	176
Butane-n	-0.5	31.1
Butyric acid n	162.5	316
Carbolic Acid (phenol)	182.2	360
Carbon Dioxide	-78.5	-109.3
Carbon Disulfide	46.2	115
Carbon Tetrachloride	76.7	170
Chloroform	62.2	142
Decane-n	173	343
Diethyl Ether	34.7	94.4
Ethane	-88	-127
Ether	35	95
Ethyl Acetate	77.2	171
Ethyl Alcohol	77.85	172.13
Ethyl Bromide	38.4	101
Ethylene Bromide	131.7	269
Ethylene Glycol	197	386
Freon refrigerant R-11	23.8	74.9
Freon refrigerant R-12	-29.8	-21.6
Freon refrigerant R-22	-41.2	-42.1
Furfurol	161.7	323
Glycerin	290	554
Glycerine	290	554
Heptane-n	98.4	209.2
Hexane-n	68.7	155.7
Jet Fuel	163	325
Linseed Oil	287	548
Methyl Acetate	57.2	135
Methyl Iodide	42.6	108
Milk	100.167	212.3
Naphthalene	218	424
Nitrobenzene	210.9	412
Nonane-n	150.7	302
Octane-n	125.6	258
Olive Oil	300	570
Pentane-n	36	96.9
Petrol	95	203

Petroleum	210	410
Phenol	182	359
Propane	-43	-45
Propionic Acid	141	286
Propylene	-47.7	-53.9
Propylene Glycol	187	368
Tar	300	572
Toluene	110.6	231
Turpentine	160	320
Water (fresh)	100	212
Xylene-o	142.7	287

Table of Specific Latent Heat of Fusion and Vaporization of Materials

Material	Latent heat of fusion (KJ/Kg)	Latent heat of vaporization (KJ/Kg)
Aluminum	397	10,900
Argon	29.5	161
Bismuth	54.0	723
Bromine (Br_2)	132	375
Chlorine (Cl_2)	181	576
Copper	209	4730
Gold	63.7	1645
Helium	3.45	20.7
Hydrogen (H_2)	59.5	445
Iron	247	6090
Krypton	16.3	108
Lead	23.0	866
Lithium	432	21,200
Mercury	11.4	295
Neon	16.8	84.8
Nickel	298	6430
Nitrogen (N_2)	25.3	199
Oxygen (O_2)	13.7	213
Plutonium (ε)	11.6	1370
Silicon	1790	12,800

Silver	105	2390
Sodium	113	4240
Sulfur	53.6	1400
Tin	59.2	2490
Titanium	296	8880
Tungsten	285	4390
Uranium	38.4	1750
Zinc	112	1890
Butane	80.2	
Carbon dioxide	571	205
Ethane	95.1	
Freon 12, 30		166.2
Freon 12,0		152.8
Freon 12, +30		136.3
Methane	58.4	112
Propane	80.1	
Water, 0	334	2501
Water, 25		2441
Water, 100		2258
Butter	60	
Oil, Olive	267	
Oil, Peanut	22	

Magnetic Permeability of Material

Material	Permeability, μ (H/m)	Relative Permeability (μ/μ_o)
Air	$1.25663753 \cdot 10^{-6}$	1.00000037
Aluminum	$1.256665 \cdot 10^{-6}$	1.000022
Austenitic stainless steel[1]	$1.260 \cdot 10^{-6}$ - $8.8 \cdot 10^{-6}$	1.003 – 7
Bismuth	$1.25643 \cdot 10^{-6}$	0.999834
Carbon Steel	$1.26 \cdot 10^{-4}$	100
Cobalt-Iron (high permeability strip material)	$2.3 \cdot 10^{-2}$	18000
Copper	$1.256629 \cdot 10^{-6}$	0.999994
Ferrite (nickel zinc)	$2.0 \cdot 10^{-5}$ – $8.0 \cdot 10^{-4}$	16 – 640

Ferritic stainless steel (annealed)	$1.26 \ 10^{-3}$ - $2.26 \ 10^{-3}$	1000 – 1800
Hydrogen	$1.2566371 \ 10^{-6}$	1
Iron (99.8% pure)	$6.3 \ 10^{-3}$	5000
Iron (99.95% pure Fe annealed in H)	$2.5 \ 10^{-1}$	200000
Martensitic stainless steel (annealed)	$9.42 \ 10^{-4}$ - $1.19 \ 10^{-3}$	750 – 950
Martensitic stainless steel (hardened)	$5.0 \ 10^{-5}$ - $1.2 \ 10^{-4}$	40 – 95
Nanoperm	$1.0 \ 10^{-1}$	80000
Neodymium magnet	$1.32 \ 10^{-6}$	1.05
Nickel	$1.26 \ 10^{-4}$ - $7.54 \ 10^{-4}$	100 – 600
Permalloy	$1.0 \ 10^{-2}$	8000
Platinum	$1.256970 \ 10^{-6}$	1.000265
Sapphire	$1.2566368 \ 10^{-6}$	0.99999976
Superconductors	0	0
Teflon	$1.2567 \ 10^{-6}$	1
Vacuum (μ_0)	1.256637×10^{-6}	**1**
Water	$1.256627 \ 10^{-6}$	0.999992
Wood	$1.25663760 \ 10^{-6}$	1.00000043

Table of Viscosity of Materials

Material	Temperature (ºC)	Viscosity (cP)
1-Propanol (propyl alcohol)		1.945
2-Propanol (isopropyl alcohol)		2.052
Acetate Glue	20	1200-1400
Acetic acid		1.056
Acetic acid		1.155
Acetone		0.302
Acetone		0.316

Alcohol, ethyl (ethanol)		1.095
Alcohol, methyl (methanol)		0.56
Alcohol, propyl		1.92
Alkyd resins	20	500-3000
Apple puree	20	1500
Baby food	40	1.4
Baby Food	93	1400
Batter	30	29500
Beet Sauce	76	1950
Benzene		0.601
Benzene		0.604
Biscuit Cream Premix	18	29200
Bone Oil	54	48
Bone oil	20	300
Brewer's yeast	20	370
Brewers Yeast	18	368
Bromine		0.944
Bromine		0.95
Broth Mix	18	430
Butter	40	30000
Butter cream, sour	20	550
Butter Deodorised	50	45
Butter Fat	65	20
Butter Fat	43	42
Butter fat (ghee)	40	45
Canola oil	30	46.2
Carbon Disulfide		0.36
Carbon Tetrachloride		0.91
Carob Bean Sauce	30	1500
Castor Oil	80	36
Castor Oil	27	580
Castor Oil		650
Castor oil	20	1000-1500
Castrol Oil		1000
Caustic soda, 50%	20	45
Chinawood Oil	21	300
Chloroform		0.53
Chocolate	49	280

Chocolate		25000
Chocolate sauce	50	280
Citrus Fruit Pulp	20	600
Cleaning emulsion	70	2420
Cocoa Butter	100	0.5
Cocoa Butter	60	50
Cocoa butter	60	50
Cocoa mass	20	4000
Coconut Oil	38	30
Coconut Oil	24	55
Coconut oil	20	60
Cod Oil	38	32
Cod-liver oil	40	35
Coffee Liquor 30-40%	20	10-100
Concentrated milk	40	80
Concentrated milk, sugared	20	6100
Condensed Milk	40-50	40-80
Condensed Milk 75% Solids	20	2160
Corn Oil	57	28
Corn oil	60	30
Cottage Cheese	18	30000
Cotton oil	20	60
Cotton Seed Oil	52	24
Cotton Seed Oil	24	62
Cream (30-50% fat content)	20	15-115
Cream 30% Fat	16	14
Cream 45% Fat	16	48
Cream 50% Fat	32	55
Cream 50% Fat	16	112
Cresol Crystals	18	10
Custard	85-90	1500
Decane		0.859
Detergents	70	1470
Diethylene	21	32
Dipropylene glycol	20	107
Dodecane		1.374
Edible Oil	20	65
Ethanol		1.074

Ether		0.223
Ethylene	21	18
Ethylene Glycol		16.2
Fruit juice	20	50
Fruit Juice	18	55-75
Fruit juice concentrate	20	2500
Fruit wort	20	600
Gear oil SAE 140	20	2700
Gear oil SAE 90	20	700
Gelatine	45	1200
Gelatine 37% Solids	43	1190
Glucose	25-30	4300-8600
Glucose	25-30	4300-6800
Glycerine		950
Glycerine 100%	38	176
Glycerine 100%	20	648
Glycerol		1.412
Glycol	20	40
Glyzerol, 100%	20	1490
Glyzerol, 100%	10	4500
Glyzerol, 100%	0	12100
Gravy	80	110
Gravy Slurry	80	110
Hand Cream	18	780
Hand crème	20	8000
Heptane		0.376
Hexane		0.297
Honey	40	2000
Honey	20	~2000-10000
Honey		10000
Hydraulic oil HLP 100	20	300
Hydraulic oil HLP 46	20	120
Hydraulic oil HLP 68	20	195
Hydrazine		0.876
Iodine pentafluoride		2.111
Isopropyl Alcohol	85	1.9
Jam	20	8500
Jam Garnish	16	8440

Karo Syrup		5000
Kerosene		1.64
Ketchup		50000
Ketchup[a]	25	~ 5000-20000
Lacquer 25% Solids	18	3000
Lacquers (25% pigments)	20	3000
Lard	38	62
Lard	40	65
Lard Oil	38	40-47
Latex Emulsion	65	48
Latex Emulsion	24	200
Latex emulsion	20	200
Linseed oil	40	30
Linseed Oil		33.1
Linseed Oil Raw	38	29
Liqueurs	20	10-100
Liquid egg	45	150
Liquid soap	60	85
Liquid wax	90	500
Lube oil	20	60-200
Machine oil, heavy	20	600
Machine oil, light	20	150
Malt extract	20	9500
Malt Extract	60	3000
Malt Extract 80%	18	9500
Mayonnaise	20	2000
Mayonnaise	20	20000
Mercury		1.526
Mercury		1.53
Methanol		0.553
Milk	52	1
Milk	18	2
Milk	20	2
Milk		3
Milk Whey 48% Sugar	40	800-1500
Mincemeat	30	100000
Molasses, 80°Brix	20	10000
Molasses, 83°Brix	20	50000

Molasses, 85°Brix	20	100000
Motor oil SAE 10	20	50
Motor oil SAE 15	20	130
Motor oil SAE 15W40	-15	3
Motor oil SAE 15W40	20	390
Motor oil SAE 5	20	30
Motor oil SAE 50	20	750
Mousse au Chocolat	40	1.5
Mousse Mix	5	1200
Mustard		70000
NaOH 20%	18	1
NaOH 30%	18	1
NaOH 40%	18	20
Octane		0.51
Oleic acid	20	40
Olive Oil	38	40
Olive oil	40	40
Olive oil	26	56.2
Orange Juice Concentrate	80	91
Orange Juice Concentrate	80	330
Orange Juice Concentrate	20	630
Orange Juice Concentrate	20	2410
Palm Oil	38	43
Palm oil	40	45
Paraffin Emulsion	18	3000
Parrafin emulsion	20	3000
Peanut Butter		250000
Peanut butter[a]		~ 10000-1000000
Peanut Oil	38	38
Peanut oil	40	40
Pectin	38	300
Pectin	40	300
Pectin	27	345
Phenol		8
Pitch	10-30 (variable)	230000000000
Plastisol	18	28000
Polyester	30	3000

Polyester resin	30	3
Polyisobutylene	85	12500
Polymer solution	20	20000
Polyol (A-component)	10	85000
Polyol, unpigmented	20	500-5000
Polypropylene	50	240000
Potassium hydroxide	20	67
Pottage	20	430
Printers Ink	54	238-660
Printers Ink	38	550-2200
Printing inks	40	550-2200
Process Cheese	80	6500
Process Cheese	18	30000
Propane		0.11
Propylene		0.09
Propylene	21	52
Propylene glycol		42
Pudding	40	1
Rape oil	20	160
Resin solution	20	7.1
Resin Solution	24	880
Resin Solution	21	975
Resin Solution	18	7140
Rice Pudding	100	10000
SAE 10 Motor Oil		85-140
SAE 20 Motor Oil		140-420
SAE 30 Motor Oil		420-650
SAE 40 Motor Oil		650-900
Salad Cream	18	1300-2600
Salad cream	20	1300-2600
Sauce – Apple	80	500
Shampoo	20	3
Shampoos	36	3000
Soap Arylan	60	630
Soap Solution	60	82
Soft cheese	60	30
Sorbitol	20	200
Sour Cream		100000

Soya Bean Oil	80	12
Soya Bean Oil	24	60
Soya bean oil	20	60
Soya bean oil, treated	20	600-800
Soya Bean Slurry	50-90	5000-10,000
Sperm Oil	38	24
Squalane		31.123
Sulphonic Acid	30	125
Sunflower oil	26	48.8
Toluene		0.55
Tomato Ketchup	30	1000
Tomato Paste 30%	18	195
Tomato puree	20	195
Tooth paste	40	70000
Toothpaste	18	70,000-100,000
Train oil	20	100
Transformer oil	20	30
Transformer oil	10	75
Treacle, 65°Brix	20	120
Treacle, 70°Brix	20	400
Triacetate Dope	40	48000-60000
Trichlorofluoromethane refrigerant R-11		0.42
Triethylene	21	40
Turbine oil	20	200-1100
Turpentine		1.375
Turpentine	16	2
Vinegar	20	12-15
Vitamine oil	10	4.5
Water		0.89
Water		1
Water varnish	20	900
Water, Fresh		0.89
Wax	93	500
Whale Oil	38	25-39
Whey	40	800-1500
Whole Egg	4.5	150

Whole milk	20	2.12
Yeast Surry	18	20
Yoghourt	40	150
Yoghurt	40	152

The viscosity is given in centiPoise (cP), which is equivalent to milliPascal seconds (mPa.s).

Table of Linear Expansivity/Coefficient of linear expansion of Materials

Metals	Coefficient of linear expansion (10^{-6} K^{-1})
Aluminium	23.1
Benzocyclobutene	42
Brass	19
Carbon steel	10.8
Concrete	12
Copper	17
Diamond	1
Ethanol	250
Gallium(III) Arsenide	5.8
Gasoline	317
Gold	14
Ice	51
Iron	11.8
Lead	29
Magnesium	26
Mercury	61
Nickel	13
Platinum	9
Water	69
Silicon	2.56
Silver	18

Table of Volume Expansivity/Coefficient of Volume Expansion of Liquids

Liquid	Volume/cubic expnsivity (/K)
Acetic acid	0.00110
Acetone	0.00143
Alcohol, ethyl (ethanol)	0.00109
Alcohol, methyl (methanol,wood alcohol, wood naphtha, wood spirits, CH_3OH)	0.00149
Ammonia	0.00245
Aniline	0.00085
Benzene	0.00125
Bromine	0.00110
Calcium Chloride, 5.8% solution	0.00025
Calcium Chloride, 40.9% solution	0.00046
Carbon disulfide	0.00119
Carbon tetrachloride	0.00122
Chloroform	0.00127
Engine oil	0.0007
Ether	0.00160
Ethyl acetate	0.00138
Ethylene glycol	0.00057
Dichlorodifluoromethane refrigerant R-12	0.0026
n-Heptane	0.00124
Hydrochloric acid, 33.2% solution	0.00046
Hydrogen (liquid 20.3 K)	0.0151
Isobutyl alcohol	0.00094
Gasoline	0.00095
Glycerine (glycerol)	0.00050
Kerosene, jet fuel	0.00099
Mercury	0.00018
Methyl iodide	0.0012
Nitrogen (liquid 100 K)	0.009
n-Octane	0.00114
Oil (unused engine oil)	0.00070
Olive oil	0.00070

Oxygen (liquid 89 K)	0.002
Paraffin oil	0.000764
Petroleum	0.0010
n-Pentane	0.00158
Phenol	0.0009
Potassium chloride, 24.3% solutiuon	0.00035
Sodium	0.00027
Sodium chloride, 20.6% solution	0.00041
Sodium sulfate, 24% solution	0.00041
Sulfuric acid, concentrated	0.00055
Toluene	0.00108
Trichloroethylene	0.001170
Turpentine	0.001000
Water at 20°C	0.000214

Table of Emf and Property of Common Cells

Cell (Common Name)	Emf	Anode	Cathode	Electrolyte
nickel-cadmium	1.2 V	Cadmium	NiO(OH)	Water, potassium hydroxide
nickel–metal hydride	1.2 V	Mischmetal (hydrogen absorbing)	Nickel	Water, potassium hydroxide
Zinc carbon	1.5 V	Zinc	Carbon, manganese dioxide	Water, ammonium or zinc chloride
Lead–acid	2.1 V	Lead	Lead dioxide	Water, sulfuric acid
Lithium-ion	3.6 V to 3.7 V	Graphite	$LiCoO_2$	Organic solvent, Li salts
Mercury cell	1.35 V	Zinc	HgO	Water, sodium or potassium hydroxide

Table of Young Modulus of Materials

Material	Young's Modulus, Y (Pa)
Iron	21×10^{10}
Nickel	21×10^{10}
Steel	20×10^{10}
Copper	11×10^{10}
Brass	9.0×10^{10}
Aluminum	7.0×10^{10}
Crown Glass	6.0×10^{10}
Cortical Bone	$7 \times 10^{9} - 30 \times 10^{9}$
Lead	1.6×10^{10}
Tendon	2×10^{7}
Rubber	$7 \times 10^{5} - 40 \times 10^{5}$
Blood vessels	2×10^{5}

Table of Shear Modulus of Materials

Material	Shear Modulus, S (Pa)
Nickel	7.8×10^{10}
Iron	7.7×10^{10}
Steel	7.5×10^{10}
Copper	4.4×10^{10}
Brass	3.5×10^{10}
Aluminum	2.5×10^{10}
Crown Glass	2.5×10^{10}
Lead	0.6×10^{10}
Rubber	$2 \times 10^{5} - 10 \times 10^{5}$

Table of Surface Tension of Liquids

Liquid	Surface Tension *(N/m)*
Acetaldehyde	0.021
Acetic acid, Ethanoic acid	0.027
Acetic anhydride, Acetyl acetate	0.032

Acetone, 2-Propanone	0.024
Acetonitrile, Methyl cyanide	0.287
Allyl alcohol	0.025
Ammonia, R-717	0.021
Aniline, Benzenamine	0.042
Anisole, Methoxybenzene	0.035
Benzene, Annulene	0.028
Benzonitrile, Phenyl cyanide	0.039
Benzylamine	0.039
Bromine	0.041
Bromobenzene	0.035
Bromoethane	0.024
n-Butane	0.023
1-Butanol, Butyl alcohol	0.025
Butyl acetate	0.025
Butylamine	0.023
Diethyl ether	0.017
Carbon dioxide	0.00056
Carbon disulfide	0.032
Carbon tetrachloride	0.027
Clorobenzene, Phenyl chloride	0.033
Chlorodifluoromethane, HCFC-22	0.008
Chloroform	0.0271
1-Chlorohexane, Hexyl chloride	0.026
1-Chloropentane	0.024
p-Cresol	0.035
Cyclohexane	0.024
Cyclohexanol	0.033
Cyclohexene	0.026
Cyclopentane	0.022
Decane	0.024
Dibutylamine	0.024
Dichlorodifluoromethane, CFC-12	0.0086
Diethylene glycol	0.045
Diethyl ether, Ethyl ether	0.017
Diethyl sulfide, Ethyl sulfide	0.025
Ethane	0.00048
Ethanol, Ethyl Alcohol, Pure Alcohol, Grain Alcohol, Drinking Alcohol	0.022

Ethanolamine, glycinol	0.048
Ethyl acetate	0.024
Ethylamine, Ethanamine	0.019
Ethylbenzene, Phenylethane	0.029
Ethyl benzoate	0.035
Ethyl bromide	0.025
Ethyl mercaptan	0.024
Ethylene glycol	0.0477
Formamide	0.057
Formixc acis, Methanoic acid	0.037
Furfural	0.043
n-Heptane	0.020
Heptanoic acid, Enanthic acid	0.028
Hexadekane, Cetane	0.027
n-Hexane	0.018
Hexanenitrile, Capronitrile	0.027
1-Hexanol, Caproyl alcohol	0.026
1-Hexene	0.018
Hydrazine	0.066
Glycerol	0.064
Isobenzene, Phenyl iodide	0.039
Isobutane, 2-Methylpropane	0.010
Isobutyl acetate. 2-Methylpropyl acetate	0.023
Isobutyric acid	0.025
Isopropanol, 2-propanol, Isopropyl Alcohol, Rubbing Alcohol, Sec-propyl Alcohol, s-Propanol	0.022
Mercury, Quicksilver	0.485
Methanol, Methyl alcohol	0.022
Methyl acetate	0.025
Methyl formate	0.025
Nitrobenzene (50°C)	0.041
Nitromethane, Nitrocarbol	0.036
Nonane	0.022
Octane	0.021
Pentane	0.015
Pentyl acetate	0.025
Propane, LPG	0.007

1-Propanol, Propyl alcohol	0.023
n-Propyl alcohol	0.024
n-Propyl benzene	0.030
Pyridine	0.037
Trichloromethane, Chloroform	0.023
Toluene, Methylbenzene	0.028
Trifluormethane, Fluoroform	0.00003
Undecane, Hendecane	0.025
Water at 20°C	0.072
Water, soapy at 20°C	0.0250 - 0.0450
Water-d_2, Heavy Water	0.071
Xenon (10°C)	0.00044
o-Xylene	0.029
m-Xylene	0.028
p-Xylene	0.028

Table of Coefficient of Static Friction of Surfaces Combination

Material	on material	Surface condition	Coefficient of static friction, μ
Aluminum	Aluminum	Clean and Dry	1.05 - 1.35
Aluminum	Aluminum	Lubricated and Greasy	0.3
Aluminum-bronze	Steel	Clean and Dry	0.45
Aluminum	Mild Steel	Clean and Dry	0.61
Aluminum	Snow	Wet 0°C	0.4
Aluminum	Snow	Dry 0°C	0.35
Brake material	Cast iron	Clean and Dry	0.4
Brake material	Cast iron (wet)	Clean and Dry	0.2
Brass	Steel	Clean and Dry	0.51
Brass	Steel	Lubricated and Greasy	0.19
Brass	Steel	Castor oil	0.11
Brass	Cast Iron	Clean and Dry	
Brick	Wood	Clean and Dry	0.6

Bronze	Steel	Lubricated and Greasy	0.16
Bronze - sintered	Steel	Lubricated and Greasy	0.13
Cadmium	Cadmium	Clean and Dry	0.5
Cadmium	Cadmium	Lubricated and Greasy	0.05
Cadmium	Chromium	Clean and Dry	0.41
Cadmium	Chromium	Lubricated and Greasy	0.34
Cast Iron	Cast Iron	Clean and Dry	1.1
Cast iron	Mild Steel	Clean and Dry	0.4
Cast iron	Mild Steel	Lubricated and Greasy	0.21
Car tire	Asphalt	Clean and Dry	0.72
Car tire	Grass	Clean and Dry	0.35
Carbon (hard)	Carbon	Clean and Dry	0.16
Carbon (hard)	Carbon	Lubricated and Greasy	0.12 - 0.14
Carbon	Steel	Clean and Dry	0.14
Carbon	Steel	Lubricated and Greasy	0.11 - 0.14
Chromium	Chromium	Clean and Dry	0.41
Chromium	Chromium	Lubricated and Greasy	0.34
Copper-Lead alloy	Steel	Clean and Dry	0.22
Copper	Copper	Clean and Dry	1.6
Copper	Copper	Lubricated and Greasy	0.08
Copper	Cast Iron	Clean and Dry	1.05
Copper	Mild Steel	Clean and Dry	0.53
Copper	Glass	Clean and Dry	0.68
Cotton	Cotton	Threads	0.3
Diamond	Diamond	Clean and Dry	0.1
Diamond	Diamond	Lubricated and Greasy	0.05 - 0.1
Diamond	Metals	Clean and Dry	0.1 - 0.15
Diamond	Metal	Lubricated and	0.1

		Greasy	
Glass	Glass	Clean and Dry	0.9 - 1.0
Glass	Glass	Lubricated and Greasy	0.1 - 0.6
Glass	Metal	Clean and Dry	0.5 - 0.7
Glass	Metal	Lubricated and Greasy	0.2 - 0.3
Glass	Nickel	Clean and Dry	0.78
Glass	Nickel	Lubricated and Greasy	0.56
Graphite	Steel	Clean and Dry	0.1
Graphite	Steel	Lubricated and Greasy	0.1
Graphite	Graphite (in vacuum)	Clean and Dry	0.5 - 0.8
Graphite	Graphite	Clean and Dry	0.1
Graphite	Graphite	Lubricated and Greasy	0.1
Hemp rope	Timber	Clean and Dry	0.5
Horseshoe	Rubber	Clean and Dry	0.68
Horseshoe	Concrete	Clean and Dry	0.58
Ice	Ice	Clean 0°C	0.1
Ice	Ice	Clean -12°C	0.3
Ice	Ice	Clean -80°C	0.5
Ice	Wood	Clean and Dry	0.05
Ice	Steel	Clean and Dry	0.03
Iron	Iron	Clean and Dry	1.0
Iron	Iron	Lubricated and Greasy	0.15 - 0.20
Leather	Oak	Parallel to grain	0.61
Leather	Metal	Clean and Dry	0.4
Leather	Metal	Lubricated and Greasy	0.2
Leather	Wood	Clean and Dry	0.3 - 0.4
Leather	Clean Metal	Clean and Dry	0.6
Leather	Cast Iron	Clean and Dry	0.6
Leather fiber	Cast iron	Clean and Dry	0.31
Leather fiber	Aluminum	Clean and Dry	0.30
Magnesium	Magnesium	Clean and Dry	0.6

Magnesium	Magnesium	Lubricated and Greasy	0.08
Masonry	Brick	Clean and Dry	0.6 - 0.7
Mica	Mica	Freshly cleaved	1.0
Nickel	Nickel	Clean and Dry	0.7 - 1.1
Nickel	Nickel	Lubricated and Greasy	0.28
Nylon	Nylon	Clean and Dry	0.15 - 0.25
Nylon	Steel	Clean and Dry	0.4
Nylon	Snow	Wet $0^{o}C$	0.4
Nylon	Snow	Dry $-10^{o}C$	0.3
Oak	Oak (parallel grain)	Clean and Dry	0.62
Oak	Oak (cross grain)	Clean and Dry	0.54
Paper	Cast Iron	Clean and Dry	0.20
Phosphor-bronze	Steel	Clean and Dry	0.35
Platinum	Platinum	Clean and Dry	1.2
Platinum	Platinum	Lubricated and Greasy	0.25
Plexiglas	Plexiglas	Clean and Dry	0.8
Plexiglas	Plexiglas	Lubricated and Greasy	0.8
Plexiglas	Steel	Clean and Dry	0.4 - 0.5
Plexiglas	Steel	Lubricated and Greasy	0.4 - 0.5
Polystyrene	Polystyrene	Clean and Dry	0.5
Polystyrene	Polystyrene	Lubricated and Greasy	0.5
Polystyrene	Steel	Clean and Dry	0.3 - 0.35
Polystyrene	Steel	Lubricated and Greasy	0.3 - 0.35
Polyethylene	Polytehylene	Clean and Dry	0.2
Polyethylene	Steel	Clean and Dry	0.2
Polyethylene	Steel	Lubricated and Greasy	0.2
Rubber	Rubber	Clean and Dry	1.16
Rubber	Cardboard	Clean and Dry	0.5 - 0.8
Rubber	Dry Asphalt	Clean and Dry	0.9
Silk	Silk	Clean	0.25

Silver	Silver	Clean and Dry	1.4
Silver	Silver	Lubricated and Greasy	0.55
Sapphire	Sapphire	Clean and Dry	0.2
Sapphire	Sapphire	Lubricated and Greasy	0.2
Silver	Silver	Clean and Dry	1.4
Silver	Silver	Lubricated and Greasy	0.55
Skin	Metals	Clean and Dry	0.8 - 1.0
Steel	Steel	Clean and Dry	0.5 - 0.8
Steel	Steel	Lubricated and Greasy	0.16
Steel	Steel	Castor oil	0.15
Steel	Steel	Light mineral oil	0.23
Steel	Steel	Lard	0.11
Steel	Graphite	Clean and Dry	0.21
Straw Fiber	Cast Iron	Clean and Dry	0.26
Straw Fiber	Aluminum	Clean and Dry	0.27
Tarred fiber	Cast Iron	Clean and Dry	0.15
Tarred fiber	Aluminum	Clean and Dry	0.18
Polytetrafluoroethylene (PTFE) (Teflon)	Polytetrafluoroethylene (PTFE)	Clean and Dry	0.04
Polytetrafluoroethylene (PTFE)	Polytetrafluoroethylene (PTFE)	Lubricated and Greasy	0.04
Polytetrafluoroethylene (PTFE)	Steel	Clean and Dry	0.05 - 0.2
Polytetrafluoroethylene (PTFE)	Snow	Wet 0°C	0.05
Polytetrafluoroethylene (PTFE)	Snow	Dry 0°C	0.02
Tungsten Carbide	Steel	Clean and Dry	0.4 - 0.6
Tungsten Carbide	Steel	Lubricated and Greasy	0.1 - 0.2
Tungsten Carbide	Tungsten Carbide	Clean and Dry	0.2 - 0.25
Tungsten Carbide	Tungsten	Lubricated and	0.12

	Carbide	Greasy	
Tungsten Carbide	Copper	Clean and Dry	0.35
Tungsten Carbide	Iron	Clean and Dry	0.8
Tire, dry	Road, dry	Clean and Dry	1
Tire, wet	Road, wet	Clean and Dry	0.2
Wax, ski	Snow	Wet 0°C	0.1
Wax, ski	Snow	Dry 0°C	0.04
Wax, ski	Snow	Dry -10°C	0.2
Wood	Clean Wood	Clean and Dry	0.25 - 0.5
Wood	Wet Wood	Clean and Dry	0.2
Wood	Clean Metal	Clean and Dry	0.2 - 0.6
Wood	Wet Metals	Clean and Dry	0.2
Wood	Stone	Clean and Dry	0.2 - 0.4
Wood	Concrete	Clean and Dry	0.62
Wood	Brick	Clean and Dry	0.6
Wood - waxed	Wet snow	Clean and Dry	0.14
Zinc	Cast Iron	Clean and Dry	0.85
Zinc	Zinc	Clean and Dry	0.6
Zinc	Zinc	Lubricated and Greasy	0.04

Coefficient of Kinetic (Sliding) Friction of Surfaces Combination

Material	on material	Surface Condition	Coefficient of kinetic friction, μ
Aluminum	Aluminum	Clean and Dry	1.4
Aluminum	Mild Steel	Clean and Dry	0.47
Brass	Steel	Clean and Dry	0.44
Brass	Cast Iron	Clean and Dry	0.3
Brass	Ice	Clean 0°C	0.02
Brass	Ice	Clean -80°C	0.15
Bronze	Cast Iron	Clean and Dry	0.22
Cadmium	Mild Steel	Clean and Dry	0.46
Cast Iron	Cast Iron	Clean and Dry	0.15
Cast Iron	Cast Iron	Clean and Dry	0.15
Cast Iron	Cast Iron	Lubricated and	0.07

		Greasy	
Cast Iron	Oak	Clean and Dry	0.49
Cast Iron	Oak	Lubricated and Greasy	0.075
Cast iron	Mild Steel	Clean and Dry	0.23
Cast iron	Mild Steel	Lubricated and Greasy	0.133
Copper	Cast Iron	Clean and Dry	0.29
Copper	Mild Steel	Clean and Dry	0.36
Copper	Mild Steel	Lubricated and Greasy	0.18
Copper	Mild Steel	Oleic acid	0.18
Copper	Glass	Clean and Dry	0.53
Garnet	Steel	Clean and Dry	0.39
Glass	Glass	Clean and Dry	0.4
Glass	Glass	Lubricated and Greasy	0.09 - 0.12
Ice	Ice	Clean 0°C	0.02
Ice	Ice	Clean -12°C	0.035
Ice	Ice	Clean -80°C	0.09
Lead	Cast Iron	Clean and Dry	0.43
Leather	Oak	Parallel to grain	0.52
Leather	Cast Iron	Clean and Dry	0.56
Magnesium	Steel	Clean and Dry	0.42
Magnesium	Cast Iron	Clean and Dry	0.25
Nickel	Nickel	Clean and Dry	0.53
Nickel	Nickel	Lubricated and Greasy	0.12
Nickel	Mild Steel	Clean and Dry	0.64
Nickel	Mild Steel	Lubricated and Greasy	0.178
Oak	Oak (parallel grain)	Clean and Dry	0.48
Oak	Oak (cross grain)	Clean and Dry	0.32
Oak	Oak (cross grain)	Lubricated and Greasy	0.072
Rubber	Dry Asphalt	Clean and Dry	0.5 - 0.8
Rubber	Wet Asphalt	Clean and Dry	0.25 - 0.75
Rubber	Dry Concrete	Clean and Dry	0.6 - 0.85

Rubber	Wet Concrete	Clean and Dry	0.45 - 0.75
Steel	Steel	Clean and Dry	0.42
Steel	Steel	Castor oil	0.081
Steel	Steel	Stearic Acid	0.15
Steel	Steel	Lard	0.084
Steel	Steel	Graphite	0.058
Polytetrafluoro ethylene (PTFE) (Teflon)	Polytetrafluoroe thylene (PTFE)	Clean and Dry	0.04
Tin	Cast Iron	Clean and Dry	0.32
Wood - waxed	Wet snow	Clean and Dry	0.1
Wood - waxed	Dry snow	Clean and Dry	0.04
Zinc	Cast Iron	Clean and Dry	0.21

Table of Triple Point Temperature of Some Materials

Substance	Triple point (K)	Triple point (°C)
Hydrogen triple point	13.8033	−259.3467
Neon triple point	24.5561	−248.5939
Oxygen triple point	54.3584	−218.7916
Argon triple point	83.8058	−189.3442
Mercury triple point	234.3156	−38.8344
Water triple point	273.16	0.01

Table of Temperature Scale and their Fixed Points

Temperature Scale	Lower fixed point	Upper fixed point
Kelvin (K)	273.15	373.1339
Celsius (°C)	0	99.9839
Fahrenheit (°F)	32	211.9710
Rankine (°R)	491.67	671.6410

Table of Vapour Pressure of Liquids

Liquid	Vapour pressure (kPa)
Acetaldehyde	120
Acetic acid	2.1
Acetic acid anhydride	0.68
Acetone	30
Allyl alcohol	2.3
Allyl chloride	40
Aluminum nitrate, 10% solution	2.4
Aluminum sulphate, 10% solution	2.4
Amyl acetate	0.47
Aniline	0.09
Beer	2.4
Benzene	14
Benzyl alcohol	0.013
Bromine	28
Butyl acetate	1.5
Butyl alcohol, 1-butanol	0.93
Butyric acid n	0.43
Calcium chloride, 25% solution	2.4
Calcium chloride, 5% solution	2.4
Carbon disulphide	48
Carbon tetrachloride	15.3
Chloroform	26
Cyclohexanol	0.9
Cyclohexanone	0.67
Ethyl acetate	14
Ethyl alcohol	12.4
Ethyl glycol	0.7
Ethylene glycol	0.007
Formic acid	5.7
Furfurol, 2-Furaldehyde	0.3
Heptane	6
Hexane	17.6
Isopropyl alcohol (rubbing alcohol)	4.4
Kerosene	0.7
Methyl acetate	28.8
Methyl alcohol, methanol	16.9
Methylene chloride,	58

dichloromethane	
Milk	2.4
Nitrobenzene	0.03
Nonane	0.6
Octane	1.9
Pentane	58
Phenol	0.05
Propanol	2.8
Propionic acid	0.47
Sea water	2.4
Sodium chloride, 25% solution	2.4
Sodium hydroxide, 20% solution	2.4
Sodium hydroxide, 30% solution	2.4
Styrene	0.85
Tetrachloroethane	0.7
Tetrachloroethylene	2.5
Toluene	3.8
Trichloroethylene	9.2
Water	2.4

Table of Absolute Refractive index of Common Materials

Material	Index of Refraction, n
Vacuum	1
Helium at 0°C and 1 atm	1.000036
Hydrogen at 0°C and 1 atm	1.000132
Air at STP	1.000277
Air at 0°C and 1 atm	1.000293
Carbon Dioxide at 0°C and 1 atm	1.001
Liquid Helium at −270°C	1.025
Ice at 0°C	1.31
Water	1.330
Acetone	1.36
Ethanol	1.361
Kerosene	1.39

Corn Oil	1.47
Glycerol	1.4729
Acrylic Glass	1.490–1.492
Benzene	1.501
Crown Glass (pure)	1.50–1.54
Plate Glass (window glass)	1.52
Sodium Chloride (table salt)	1.544
Amber	1.55
Polycabonate	1.60
Flint Glass (pure)	1.60–1.62
Bromine	1.661
Sapphire	1.762–1.778
Cubic Zirconia	2.15–2.18
Diamond	2.417
Silicon	3.42–3.48
Germanium	4.05–4.01

Table of Wavelength, Frequency and Energy of Regions of the Electromagnetic Spectrum

Component	Wavelength (m)	Frequency (Hz)	Energy (J)
Radio	$> 1 \times 10^{-1}$	$< 3 \times 10^{9}$	$< 2 \times 10^{-24}$
Microwave	$1 \times 10^{-3} - 1 \times 10^{-1}$	$3 \times 10^{9} - 3 \times 10^{11}$	$2 \times 10^{-24} - 2 \times 10^{-22}$
Infrared	$7 \times 10^{-7} - 1 \times 10^{-3}$	$3 \times 10^{11} - 4 \times 10^{14}$	$2 \times 10^{-22} - 3 \times 10^{-19}$
Visible Light	$4 \times 10^{-7} - 7 \times 10^{-7}$	$4 \times 10^{14} - 7.5 \times 10^{14}$	$3 \times 10^{-19} - 5 \times 10^{-19}$
UV	$1 \times 10^{-8} - 4 \times 10^{-7}$	$7.5 \times 10^{14} - 3 \times 10^{16}$	$5 \times 10^{-19} - 2 \times 10^{-17}$
X-ray	$1 \times 10^{-11} - 1 \times 10^{-8}$	$3 \times 10^{16} - 3 \times 10^{19}$	$2 \times 10^{-17} - 2 \times 10^{-14}$
Gamma-ray	$< 1 \times 10^{-11}$	$> 3 \times 10^{19}$	$> 2 \times 10^{-14}$

Table of Wavelength of the Colours of White Light

Colour	Wavelength (nm)	Frequency (x 10^{14}Hz)	Energy (eV)
red (limit)	700	4.29	1.77
red	650	4.62	1.91

orange	600	5.00	2.06
yellow	580	5.16	2.14
green	550	5.45	2.25
cyan	500	5.99	2.48
blue	450	6.66	2.75
violet (limit)	400	7.50	3.10

Note that the values given above are typical values only.

Table of Speed of Sounds in Some Materials

Material	Speed of sound (ms^{-1})
Air at 0°C	330
Air at 20°C	343
Aluminium	6300
Alumina Oxide	9900
Beryllium	12900
Boron Carbide	11000
Brass	4300
Cadmium	2800
Copper	4700
Diamond	12000
Glass(crown)	5300
Glycerine	1900
Gold	3200
Ice	4000
Inconel	5700
Iron	5900
Iron (cast)	4600
Lead	2200
Magnesium	5800
Mercury	1400
Molybdenum	6300
Monel	5400
Neoprene	1600
Nickel	5600
Nylon, 6.6	2600
Oil (SAE 30)	1700
Platinum	3300

Plexiglass	1700
Polyethylene	1900
Polystyrene	2400
Polyurethane	1900
Quartz	5800
Rubber, Butyl	1800
Silver	3600
Steel, Mild	5920
Steel, Stainless	5800
Teflon	1400
Tin	3300
Titanium	6100
Tungsten	5200
Uranium	3400
Water	1480
Zinc	4200

Table of Colour Code for Resistors

Color	Digit	Multiplier	Tolerance (%)
Black	0	$10^0 (1)$	
Brown	1	10^1	1
Red	2	10^2	2
Orange	3	10^3	
Yellow	4	10^4	
Green	5	10^5	0.5
Blue	6	10^6	0.25
Violet	7	10^7	0.1
Grey	8	10^8	
White	9	10^9	
Gold		10^{-1}	5
Silver		10^{-2}	10
(none)			20

Table of Half-Life of Elements

Element	Half-life
Holmium-166m	1,200 years
Berkelium-247	1,380 years
Radium-226	1,600 years
Molybdenum-93	4,000 years
Holmium-153	4,570 years
Curium-246	4,730 years
Carbon-14	5,730 years
Plutonium-240	6,563 years
Thorium-229	7,340 years
Americium-243	7,370 years
Curium-245	8,500 years
Curium-250	9,000 years
Tin-126	10,000 years
Niobium-94	20,300 years
Plutonium-239	24,110 years
Protactinium-231	32,760 years
Lead-202	52,500 years
Lanthanum-137	60,000 years
Thorium-230	75,380 years
Nickel-59	76,000 years
Thorium-230	77,000 years
Calcium-41	103,000 years
Neptunium-236	154,000 years
Uranium-233	159,200 years
Rhenium-186m	200,000 years
Technetium-99	211,000 years
Krypton-81	229,000 years
Uranium-234	245,500 years
Chlorine-36	301,000 years
Curium-248	340,000 years
Bismuth-208	368,000 years
Plutonium-242	373,300 years
Aluminum-26	717,000 years
Selenium-79	1,130,000 years
Iron-60	1,500,000 years
Beryllium-10	1,510,000 years
Zircon-93	1,530,000 years

Curium-247	1,560,000 years
Gadolinium-150	1,790,000 years
Neptunium-237	2,144,000 years
Cesium-135	2,300,000 years
Technetium-97	2,600,000 years
Dysprosium-154	3,000,000 years
Bismuth-210m	3,040,000 years
Manganese-53	3,740,000 years
Technetium-98	4,200,000 years
Palladium-107	6,500,000 years
Hafnium-182	9,000,000 years
Lead-205	15,300,000 years
Curium-247	15,600,000 years
Iodine-129	17,000,000 years
Uranium-236	23,420,000 years
Niobium-92	34,700,000 years
Plutonium-244	80,800,000 years
Samarium-146	103,000,000 years
Uranium-236	234,200,000 years
Uranium-235	703,800,000 years
Potassium-40	1,280,000,000 years
Uranium-238	4,468,000,000 years
Rubidium-87	4,750,000,000 years
Thorium-232	14,100,000,000 years
Lutetium-176	37,800,000,000 years
Rhenium-187	43,500,000,000 years
Lanthanum-138	105,000,000,000 years
Samarium-147	106,000,000,000 years
Platinum-190	650,000,000,000 years
Tellurium-123	$>1 \times 10^{13}$ years
Osmium-184	$>5.6 \times 10^{13}$ years
Gadolinium-152	1.08×10^{14} years
Tantalum-180m	$>1.2 \times 10^{15}$ years
Xenon-124	$>1.6 \times 10^{14}$ years
Indium-115	4.41×10^{14} years
Zinc-70	$>5 \times 10^{14}$ years
Hafnium-174	2.0×10^{15} years
Osmium-186	2.0×10^{15} years
Samarium-149	$>2 \times 10^{15}$ years

Neodymium-144	2.29×10^{15} years
Samarium-148	7×10^{15} years
Cadmium-113	7.7×10^{15} years
Cerium-142	$>5 \times 10^{16}$ years
Tungsten-183	$>1.1 \times 10^{17}$ years
Vanadium-50	1.4×10^{17} years
Lead-204	1.4×10^{17} years
Chromium-50	$>1.8 \times 10^{17}$ years
Tungsten-184	$>3 \times 10^{17}$ years
Calcium-48	$>6.3 \times 10^{18}$ years
Molybdenum-100	1.0×10^{19} years
Neodynium-150	$>1.1 \times 10^{19}$ years
Zircon-96	$>3.8 \times 10^{19}$ years
Selenium-82	1.1×10^{20} years
Tellurium-130	7.9×10^{20} years
Xenon-136	$>2.4 \times 10^{21}$ years
Tellurium-128	2.2×10^{24} years
Promethium-145	17.4 years
Polonium-209	102 years
Astatine-210	8.1 hours
Radon-222	3.82 days
Francium-223	22 minutes
Actinium-227	21.77 years
Californium-251	898 years
Einsteinium-252	471.7 days
Fermium-257	100.5 days
Mendelevium-258	51.5 days
Nobelium-259	58 minutes
Lawrencium-262	4 hours
Rutherfordium-265	13 hours
Dubnium-268	32 hours
Seaborgium-271	2.4 minutes
Bohrium-267	17 seconds
Hassium-269	9.7 seconds
Meitnerium-276	0.72 seconds
Darmstadtium-281	11.1 seconds
Roentgenium-281	26 seconds
Copernicium-285	29 seconds
Nihonium-284	0.48 seconds

Flerovium-289	2.65 seconds
Moscovium-289	87 milliseconds
Livermorium-293	61 milliseconds
Oganesson-294	1.8 milliseconds

Table of Binding Energy Per Nucleon for some Common Elements

Element	Mass of nucleons (u)	Nuclear mass (u)	Binding energy (MeV)	Binding energy per nucleon (MeV)
Deuterium-2	2.01594	2.01355	2.23	1.12
Helium-4	4.03188	4.00151	28.29	7.07
Lithium-7	7.05649	7.01336	40.15	5.74
Beryllium-9	9.07243	9.00999	58.13	6.46
Iron-56	56.44913	55.92069	492.24	8.79
Silver-107	107.86187	106.87934	915.23	8.55
Iodine-127	128.02684	126.87544	1072.53	8.45
Lead-206	207.67109	205.92952	1622.27	7.88
Polonium-210	211.70297	209.93683	1645.16	7.83
Uranium-235	236.90849	234.99351	1783.80	7.59
Uranium-238	239.93448	238.00037	1801.63	7.57

Table of Work Function of Metals for Photoelectric Effects

Element	Work Function, ϕ (eV)
Aluminum	4.08
Beryllium	5.0
Cadmium	4.07
Calcium	2.9
Carbon	4.81
Cesium	2.1
Cobalt	5.0
Copper	4.7
Gold	5.1
Iron	4.5

Lead	4.14
Magnesium	3.68
Mercury	4.5
Nickel	5.01
Niobium	4.3
Potassium	2.3, 2.29
Platinum	6.35
Selenium	5.11
Silver	4.26-4.73
Sodium	2.28, 2.36
Uranium	3.6
Zinc	4.3
Titanium	4.33
Vanadium	4.3
Chromium	4.5
Manganese	4.1

Table of Electrochemical Equivalent of Some Elements

Element	Electrochemical equivalent (Kg/C)
Al (Aluminum)	4.3
Ti (Titanium)	4.33
V (Vanadium)	4.3
Cr (Chromium)	4.5
Mn (Manganese)	4.1
Fe (Iron)	4.7
Co (Cobalt)	5
Ni (Nickel)	5.15
Nb (Niobium)	4.3

Table of Standard Electrode Potential of Elements

Cathode (Reduction) Half-Reaction	Standard Electrode Potential, E° (volts)
$Li^+(aq) + e^- \rightarrow Li(s)$	-3.04

$K^+(aq) + e^- \rightarrow K(s)$	-2.92
$Ca^{2+}(aq) + 2e^- \rightarrow Ca(s)$	-2.76
$Na^+(aq) + e^- \rightarrow Na(s)$	-2.71
$Mg^{2+}(aq) + 2e^- \rightarrow Mg(s)$	-2.38
$Al^{3+}(aq) + 3e^- \rightarrow Al(s)$	-1.66
$2H_2O(l) + 2e^- \rightarrow H_2(g) + 2OH^-(aq)$	-0.83
$Zn^{2+}(aq) + 2e^- \rightarrow Zn(s)$	-0.76
$Cr^{3+}(aq) + 3e^- \rightarrow Cr(s)$	-0.74
$Fe^{2+}(aq) + 2e^- \rightarrow Fe(s)$	-0.41
$Cd^{2+}(aq) + 2e^- \rightarrow Cd(s)$	-0.40
$Ni^{2+}(aq) + 2e^- \rightarrow Ni(s)$	-0.23
$Sn^{2+}(aq) + 2e^- \rightarrow Sn(s)$	-0.14
$Pb^{2+}(aq) + 2e^- \rightarrow Pb(s)$	-0.13
$Fe^{3+}(aq) + 3e^- \rightarrow Fe(s)$	-0.04
$2H^+(aq) + 2e^- \rightarrow H_2(g)$	0.00
$Sn^{4+}(aq) + 2e^- \rightarrow Sn^{2+}(aq)$	0.15
$Cu^{2+}(aq) + e^- \rightarrow Cu^+(aq)$	0.16
$ClO_4^-(aq) + H_2O(l) + 2e^- \rightarrow ClO_3^-(aq) + 2OH^-(aq)$	0.17
$AgCl(s) + e^- \rightarrow Ag(s) + Cl^-(aq)$	0.22
$Cu^{2+}(aq) + 2e^- \rightarrow Cu(s)$	0.34
$ClO_3^-(aq) + H_2O(l) + 2e^- \rightarrow ClO_2^-(aq) + 2OH^-(aq)$	0.35
$IO^-(aq) + H_2O(l) + 2e^- \rightarrow I^-(aq) + 2OH^-(aq)$	0.49
$Cu^+(aq) + e^- \rightarrow Cu(s)$	0.52
$I_2(s) + 2e^- \rightarrow 2I^-(aq)$	0.54
$ClO_2^-(aq) + H_2O(l) + 2e^- \rightarrow ClO^-(aq) + 2OH^-(aq)$	0.59
$Fe^{3+}(aq) + e^- \rightarrow Fe^{2+}(aq)$	0.77
$Hg_2^{2+}(aq) + 2e^- \rightarrow 2Hg(l)$	0.80
$Ag^+(aq) + e^- \rightarrow Ag(s)$	0.80
$Hg^{2+}(aq) + 2e^- \rightarrow Hg(l)$	0.85
$ClO^-(aq) + H_2O(l) + 2e^- \rightarrow Cl^-(aq) + 2OH^-(aq)$	0.90
$2Hg^{2+}(aq) + 2e^- \rightarrow Hg_2^{2+}(aq)$	0.90
$NO_3^-(aq) + 4H^+(aq) + 3e^- \rightarrow NO(g) + 2H_2O(l)$	0.96
$Br_2(l) + 2e^- \rightarrow 2Br^-(aq)$	1.07

$O_2(g) + 4H^+(aq) + 4e^- \rightarrow 2H_2O(l)$	1.23
$Cr_2O_7^{2-}(aq) + 14H^+(aq) + 6e^- \rightarrow$ $2Cr^{3+}(aq) + 7H_2O(l)$	1.33
$Cl_2(g) + 2e^- \rightarrow 2Cl^-(aq)$	1.36
$Ce^{4+}(aq) + e^- \rightarrow Ce^{3+}(aq)$	1.44
$MnO_4^-(aq) + 8H^+(aq) + 5e^- \rightarrow$ $Mn^{2+}(aq) + 4H_2O(l)$	1.49
$H_2O_2(aq) + 2H^+(aq) + 2e^- \rightarrow 2H_2O(l)$	1.78
$Co^{3+}(aq) + e^- \rightarrow Co^{2+}(aq)$	1.82
$S_2O_8^{2-}(aq) + 2e^- \rightarrow 2SO_4^{2-}(aq)$	2.01
$O_3(g) + 2H^+(aq) + 2e^- \rightarrow O_2(g) + H_2O(l)$	2.07
$F_2(g) + 2e^- \rightarrow 2F^-(aq)$	2.87

Table of Energy Band Gap of Semi-Conductors

Material	Symbol	Band gap (eV)
Silicon	Si	1.11
Germanium	Ge	0.66
Indium Antimonide	InSb	0.17
Indium Arsenide	InAs	0.36
Indium Phosphide	InP	1.27
Gallium Phosphide	GaP	2.25
Gallium Arsenide	GaAs	1.43
Gallium Antimonide	GaSb	0.68
Cadmium Selenide	CdSe	1.74
Cadmium Telluride	CdTe	1.44
Zinc Oxide	ZnO	3.2
Zinc Sulphide	ZnS	3.6
Aluminium nitride	AlN	6.0
Diamond	C	5.5
Gallium nitride	GaN	3.4
Silicon nitride	Si_3N_4	5
Lead(II) sulfide	PbS	0.37
Silicon dioxide	SiO_2	9
Copper(I) oxide	Cu_2O	2.1

Table of Threshold Frequency of Metals

Metal	Threshold frequency (Hz)
Aluminum	9.846×10^{14}
Lead	9.99×10^{14}
Zinc	1.038×10^{14}
Iron	$1.086 \times x \times 10^{14}$
Copper	1.134×10^{15}
Silver	1.141×10^{15}
Nickel	1.209×10^{15}
Gold	1.231×10^{14}
Platinum	1.532×10^{15}
Calcium	6.94×10^{14}
Sodium	6.65×10^{14}
Thorium	8.22×10^{14}
Potassium	5.56×10^{14}
Caesium	4.71×10^{14}
Lithium	7.01×10^{14}
Chromium	10.9×10^{14}

In order to see other mathematics, physics and chemistry books written by the author, visit: amazon.com/author/kingzohb2. Also, you can simply go to amazon.com and search for the author's name, Kingsley Augustine, the books written by the author will show up.

If you have any enquiries, suggestions or information concerning this book, please contact the author through the email below.

Kingsley Augustine

kingzohb2@yahoo.com

Twitter handle: @kingzohb2